The Spirit
of Light Cubit

The Measure of Humanity and Spirit

Donald B. Carroll

outskirts
press

The Spirit of Light Cubit
The Measure of Humanity and Spirit
All Rights Reserved.
Copyright © 2020 Donald B. Carroll
v2.0

The opinions expressed in this manuscript are solely the opinions of the author and do not represent the opinions or thoughts of the publisher. The author has represented and warranted full ownership and/or legal right to publish all the materials in this book.

This book may not be reproduced, transmitted, or stored in whole or in part by any means, including graphic, electronic, or mechanical without the express written consent of the publisher except in the case of brief quotations embodied in critical articles and reviews.

Outskirts Press, Inc.
http://www.outskirtspress.com

Paperback ISBN: 978-1-9772-2524-5
Hardback ISBN: 978-1-9772-2527-6

Library of Congress Control Number: 2020905116

Cover Photo © Public Domain Image. All rights reserved - used with permission.

Outskirts Press and the "OP" logo are trademarks belonging to Outskirts Press, Inc.

PRINTED IN THE UNITED STATES OF AMERICA

Dedication

To my children, Jeremy, Chase, and Corwin who taught me so much; to my brother Chris, for reminding me to write in my own voice; to Alison who years ago encouraged me to write about my research and to all who taught me so much on my journey, I thank you.

Table of Contents

Chapter 1: Our Journey: Past, Present, and Future 1

Chapter 2: The map from Egypt to around the world and through time 12

Chapter 3: Humanity's urge to measure and incorporate space and time in their consciousness — squaring the circle 22

Chapter 4: Sacred structures connecting Heaven and Earth; the measurement results 42

Chapter 5: The spine symbolized as a serpent— transitioning from the physical to the spiritual 100

Chapter 6: What does this mean for ancient societies? What happened? 136

Chapter 7: What is the meaning, the message, for today? 189

Epilogue 195

Post Script 208

Book Figures 209

Recommended Reading 213

Author Bio 215

Endnotes 216

1

Our Journey: Past, Present, and Future

*"I think, I think I am, therefore I am, I think…
I'm more than that, I know I am, at least, I
think I must be. … And keep on thinking free."*
—The Moody Blues: *On the Threshold of a Dream*:
In the Beginning. 1969

Hello. Thank you for reading; I have discovered an incredible and elegant link from the past, a global link that is as integral to our journey today as it was thousands of years ago. It is a link through an ancient unit of measurement, used at sacred sites to unify, within ourselves, Heaven and Earth along with time and space; unifying the physical and spiritual. It is a symbolic message that has been left for all of us. It is as important today as it was then. It is an ancient unit of measure of 27.5 inches (70 cm). I call this measurement the Spirit of Light Cubit or spirit/light cubit.

This discovery is an epiphany-like milestone for me in my own journey through life, a journey that we all share; a journey in a physical world where we have non-physical thoughts, feelings, emotions, and consciousness. I would hazard to write that we have all looked around our world and our lives and have said more than once, "I am more than that. I think I must be." The funny thing about epiphanies, they seem to be "aha" moments that occur in a flash. I have come to believe they are more like a slow, steady, gradual buildup of questions and thoughts that reach a critical mass which suddenly breaks through. As an example, think about how water changes phases from ice through liquid to gas. Raising water's temperature between phases goes placidly along at one degree C per one calorie of heat; then, when it reaches a phase change, ice to water, then water to gas, that one degree from zero degrees C to one degree C and 99 degrees C to 100 degrees C, suddenly it takes not one calorie; it takes 80 calories at the freezing point and 540 calories at the boiling point (almost seven times more) respectively. It is those phase changes from solid to liquid and liquid to gas that take so much extra energy. I see my epiphany following the same path of phase changes; my journey through life seeking to understand both science and spirit and find reconciliation between them. Slowly did it rise and build up, seeming to stall, to plateau, but though sometimes feeling mired, I continued to put my energy into my journey. Suddenly I would break through and have such "aha" moments. These are phase changes of thought in my life. They seem to coming out of nowhere while in actuality they are a product of persistent seeking, learning and "keep on thinking".

I share with you in this first chapter a brief synopsis of a life that brought me to this moment of discovery along with a brief summation of *The Spirit of Light Cubit*'s purpose, its possible originations, and sacred sites where its use can be found, from my own hypothesis and what the archeologists, scholars, and researchers have found. The chapters following go into more detail, depth, and breadth in all these aspects.

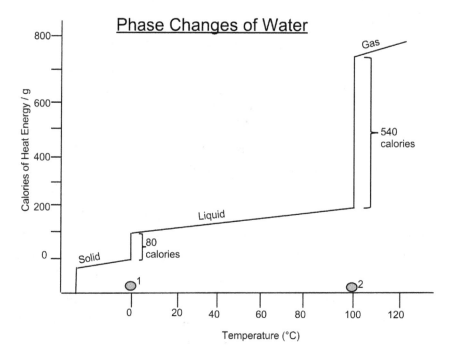

Fig. 1

From the song quote earlier you can see I am a child of the sixties—a time remembered and considered by many as a time of rebellion, cultural upheaval, questioning authority, questioning the status quo, questioning war, seeking answers outside of our parents and societal standards, seeking answers outside the mainstream religions found in our local places of worship. A time for all of us growing up in it that we thought was unique to ourselves and our generation, trying to discover who we are, why we are, what our purpose of being is; looking around ourselves, our parents, our communities, our states, our religions, and always thinking: *I'm more than that, I know I am, at least, I think I must be.*

As unique as we thought growing up in the sixties was, and it was, it is also a reiteration of every generation of human-kind asking the same questions and seeking the answers. Wonderfully every generation and every individual is magnificently unique and also beautifully connected in a sharing and Oneness; this is a my-your-ours journey. As another put it so well: To know yourself to be yourself, yet one with the whole." [1]

This is my continuing journey, a single thread woven into the tapestry of Humanity's journey. This journey in consciousness from my youth to the present, one that is still unfolding. Probably like many of you I was raised with science and religion being put in distinct categories that did not intersect. Not only did the U.S. Constitution separate church and state; so did the community I grew up in, as did most of the world since the "Age of Enlightenment." Basically, science was grounded on the scientific method, and religion was centered on faith; each stayed in their lanes. The quote "Render therefore unto Caesar the things which are Caesar's; and unto God the things that are God's" (Matthew 22:21 KJV) seems to confirm that. It is only recently I have begun to understand it doesn't have to mean separation and can be a synergistic collaboration within a greater Oneness. As for the concept of spirituality sans religion, spirituality was not even a term in my vocabulary. I vaguely remember in 1966 the *Time* magazine cover (04/08/66) "Is God Dead?" Yes, it made me pause and feel somewhat confused; it seemed like an impossible, nonsensical question in my young mind and upbringing. At this point in my life society had done a good enough job of keeping science in schools and religions in houses of worship, and never the twain shall meet. I didn't think about it much. I didn't want to think about it much—being a young teen I had other immediate concerns.

This was my consciousness, my life, until I was about sixteen. The late sixties opened up so many avenues of thought for me and more questions: Eastern religions and philosophies, space travel, war and protest against war, assassinations, drugs, "the Red scare," civil rights, countercultures. I was awash in this mind-bending potpourri of life, and experiences being stirred in at the same time of my burgeoning adulthood. All this echoed and amplified in the music of the times, not made just for dancing but for thinking and seeking: Bob Dylan's *Hard Rain*, the Beatles' "Day in the Life," the Who's *Tommy*, the Doors' *Soft Parade*, and of course the Moody Blues. The beginning of my consciousness wondering who I was, where did I fit, what it was all about (not Alfie).

This intoxicating, heady, and sometimes overwhelming brew was the crucible and genesis in which my "self" began to be formed, and in all this society and my generation wanted me to choose sides: for society, go to school, go to church on Sundays, get a good job, raise a family, fit in; for my generation it was that famous or infamous misunderstood saying, "Turn on, tune in, drop out." [2] I felt adrift and tossed about in the middle of it all. I questioned, pushed the envelope, and thought why couldn't I do both? Another aspect of desiring this culture and counterculture hybrid included my thoughts of why religion (again, I hadn't even begun to think of terms of spirituality) and science couldn't cooperate with each other toward the same goal of what our purpose is, who we are, what is consciousness. I wasn't even aware at the time that Einstein spoke along these lines: "Science without religion is lame; religion without science is blind."[3] My experience then was that neither society's maturity nor mine was ready for such a solution. I was barely aware that I was seeking religious (spiritual) answers, scientific answers, and answers about myself and existence. It was a start.

He's not the kind you have to wind up on Sundays.
—"Wind Up" by Jethro Tull, 1971

I tried to straddle the extremes of both worlds with mixed success; a nice way to write of having many bumps, bruises and failures in my journey, along with many wonderful times, some appearing spiritually linked, some not. Seeking knowledge and learning on many levels is important to the journey, but it is experience that transforms such into wisdom, even if it was a lens I saw darkly through. I always seemed to be trying to reconcile science with spirit, my life in a physical material world with one of greater consciousness and infiniteness that was beyond any physicality, and often I would feel that perhaps they were irreconcilable—two steps forward, one step back—yet I did move forward in this journey. My struggle for this reconciliation, this search for wholeness, feeling something missing, was well described by Aldous Huxley, whose writings I found years into my journey and gave me a wry smile as I thought, *Why didn't I find these earlier.* My wry smile was because I answered myself: I wasn't ready. I share a bit of Mr. Huxley's writing below.

> *We have to combine these things: to walk on this tightrope, to gather the data of perception, to be able to analyze it in terms of language, at the same time to be able to drop the language and to go into the experience.*[4]

> *If one is not oneself a sage or saint, the best thing one can do, in the field of metaphysics, is to study the works of those who were, and who because they had modified their merely human mode of being, were capable of a more than merely human kind and amount of knowledge.*[5]

Though I hadn't discovered Aldous Huxley's writings, other than his *Brave New World*, I had been casting about reading many authors and their views of metaphysics using their knowledge as guide posts on my road of experiences, and as a young man I dove into experiences. It has taken years but I have begun to see the light within myself, to begin to live beyond a "merely human mode of being." Please, do not misunderstand me. This is not a disparaging comment toward being a human being; it is remembering that we are all more than just our physicality, that our consciousness (or spirit) goes beyond our physicality. As it has been put, we are not physical beings having a spiritual experience; we are spiritual (consciousness) beings having a physical experience.[6]

This has been a personal journey for me. I think it is a personal journey for everyone and at the same time it is a journey for Humanity as a whole. The highs and lows, joys and sorrows, bumps and bruises—looking back, I am appreciative for it all, learning and growing from my life and ideally not having to repeat some of the experiences for not transforming them into present wisdom. Hopefully, at least now, I'm up to three steps forward and one step back.

At some level of consciousness I seemed to be drawn to jobs of service, of challenge and risk, intimately entwined with life and physicality. I was an ocean lifeguard for five summers, a law enforcement officer for three and a half years, and finally alit in a career as a firefighter and paramedic, rising in the ranks over a 30+year career to a chief. As I look back this was a fitting career of service, combining the hard sciences involved in firefighting and advanced emergency medical treatment with its necessity for knowledgeable, rapid decision making and acting upon it with life and property in the balance. Those of us in such service are often dealing with those in need, in highly emotionally agitated states, in danger, suffering an injury or loss and sometimes dying. Such a career made me think about the deeper and philosophical questions of life and actively search for answers.

A key aspect to a more organized, conscious growth in my seeking and journey has been learning and practicing daily meditation. I have been a daily practitioner of meditation for the last 17 years as of this writing, and it has been a wonderful boon for my life; in better understanding and appreciating my past, my present, and navigating what is to come. Some may say this is merely an individual, personal, anecdotal experience that is not measurable, and to a certain extent that is true. At the same time modern science and technology are catching up to the practice and experience of meditation. Less than 100 years ago, science basically considered the practice of meditation an esoteric religious practice, then put it in that separate category of non-scientific and non-measurable. Now as scientific and medically technology has advanced where one's heart (just the physical one), blood pressure, brain waves monitoring, PET scans and CAT scans are possible, early studies are showing the positive changes meditation can bring about that are physical and measurable. Positive results have found using mindfulness practices (consider the term "mindfulness" a synonym for meditation, used to remove any religious bias) for stress, anxiety, sleep, well-being, PTSD, weight management, and drug addiction, along with actual physical changes to the brain. Numerous corporations and the military are now offering mindfulness classes to their people.

Meditation, under many names, has been used for thousands upon thousands of years and was understood as a path to spirituality, enlightenment, and higher consciousness. I am confident its physical benefits were known in these ancient practices. I understand why modern science held it away at arm's length until they had technology to measure the effects, but at the same time it doesn't change the fact that those measurable results have been occurring already for millennia. I can't help but find some irony in that the modern schism between science and religion (spirituality) was called the "Age of Enlightenment." Again I also appreciate the causes of this schism, but it is important to remember in all this, technology only measures; human beings feel and have consciousness.

All this brings me back to my journey with the coalescing of my intuition and researched evidence aiding me toward my own conclusion that my reason and purpose for existence in physicality and projecting from that conclusion that this is our collective journey was to evolve in consciousness, ultimately to a shared universal consciousness. Many scholars and philosophers agree this is our purpose; that as science has identified a physical evolution in humans there is a concurrent consciousness evolution occurring that goes beyond our physicality that this is the journey for all of us.

Teilhard de Chardin, a geologist and paleontologist, was involved in the discovery of the human physical evolution link "Peking Man." He was also a philosopher and Jesuit priest. Along with his work in physical evolution, he wrote of the evolution of human consciousness; a consciousness he saw evolving first toward a planetary wide consciousness (Noosphere) and ultimately to a universal consciousness (Omega point). Aldous Huxley saw the evolution of human consciousness in a similar fashion:

> *That it is possible for human beings to love, know and, from virtually, to become actually identical with the Divine Ground. That to achieve this unitive knowledge of the God-head is the final end and purpose of human existence.*[7]

I can only provide the compelling evidence I have found and share of my own ongoing transformative experience. It has been an exciting, quiet, passionate, peaceful, joyous, and challenging journey—a journey that reconciles all these seemingly incompatible feelings and experiences into a more unified understanding and wisdom. It is as if I am a novice conductor of an orchestra who is beginning to learn how to blend all this cacophony into a harmonious, flowing symphony. This symphony could be called *The Ode to Oneness*. I hope to share my journey of discovery, yet let the "music" speak for itself.

The Oneness I speak of is the same as described earlier from Huxley and de Chardin. This is an apotheosis; yes, the potential deification that is the journey and purpose of all of us. Some have called it becoming co-creators or friends of God. Please, before anyone cries blasphemy, just pause for a moment and continue reading.

In the Bible's Old Testament, Psalm 82 (KJV) states, "I have said, ye are gods; and all of you are children of the most High." In the New Testament (KJV) 6 John 10:43, Jesus answered them, "Is it not written in your law, I said, Ye are gods?" Then in John 14:11, "Believe me when I say that I am in the Father and the Father is in me; or at least believe on the evidence of the works themselves. Very truly I tell you, whoever believes in me will do the works I have been doing, and they will do even greater things than these, because I am going to the Father." This is followed up with John 15:15: "No longer do I call you servants, for a servant does not know what his master is doing; but I have called you friends, for all things that I heard from My Father I have made known to you."

That is just a small sampling from the Christian side of such theology. As we are taught and as we learn and raise our consciousness, we begin to claim our spiritual birthright to become deified co-creators, no longer servants but friends of God, companions and peers. I think of it this way, being "gods and all of you are children of the most High." And from that spiritual premise, once we learn what is needed, we become friends of God. My analogy is my own family and children. I have always loved them dearly, but when they were small children they couldn't really be my friends. Now that they have learned, had their own experiences, and grown up, becoming adults, I still love them as dearly as I did when they were children, but now we are also friends. Basically God has seven billion children that He/She is waiting to grow up and become friends, peers.

Eastern traditions have similar themes: to reach spiritual enlightenment, Buddhahood, to become a Bodhisattva and help others while on Earth. None of this is blasphemous; it is the ultimate journey. Even if you go to Washington, D.C. and look up inside the capitol rotunda, you will see the work entitled "Apotheosis of Washington."

Ancient cultures understood this journey and with both their science and spirituality built "Cosmic Engines" to facilitate the completion of this journey. The mostly intact remains of some of these erected cosmic engines are still with us today, a testimony to these builders' science, engineering, architectural, and spiritual sophistication.

Fig. 2 Apotheosis of Washington: Rotunda of U.S. Capitol

In my first book, *Sacred Geometry and Spiritual Symbolism: The Blueprint for Creation*, I cast my net far and wide, sharing spiritual symbolism through the ages that had the same spiritual message to guide in the physical and spiritual unification within ourselves along with multiple avenues and evidence beyond; a net that was perhaps too wide with the potential of getting lost in too much information. Frankly I was surprised when it was agreed to be published and tried to put too many trains of thoughts in at once, perhaps clouding its purpose. Here my efforts are aimed at distilling and refining the purpose of our spiritual and physical existence and balancing the draw and fascination in such a gestalt-like journey, focusing on ancient sacred sites that exhibit compelling evidence of a shared unit of measurement and reason.

2

The map from Egypt to around the world and through time

This chapter will be a brief overview of these efforts with the following chapters going into greater detail and depth of the aspects presented.

Part of the beginning, for me, was with ancient Egypt. Today Egyptologists estimate that about two-thirds of ancient Egypt is still hidden beneath the sands. In researching the work done over the ages by a large array of explorers and Egyptologists, I felt it was the same case in finding their results; there is so much to all that has been found already and written about. Rather than digging through the sands of time in Egypt, I was digging through the dusty archives on Egypt and ancient measurements, their importance, derivations, and symbolism. I will start with Sir Flinders Petrie, considered the father of modern Egyptology, who brought detailed scientific methods and recording into the field. Sir Petrie states: *The study of ancient measures used in a country is a basis of discovering the movements of civilization between countries.*[8]

As Sir Petrie lays out the importance of shared measurements as indicators of shared communications and connections between different cultures, Iain Morley and Colin Renfrew, both acknowledged experts in archeology and Fellows of the McDonald Institute for Archeological Research, write that measurement systems go further than communication and thought processes to include metaphysical belief systems.

> *Measurement systems have provided the structure for addressing key concerns of cosmological belief systems, as well as the means for articulating relationships between human form, human action, and the world—and new understanding of relationships between events in the terrestrial world and beyond.*[9]

Presented between these covers is the evidence of Sir Petrie's statement; written here is the recognition and demonstration of an ancient unit of measurement from such a civilization with its sharing and movements among multiple peoples across the globe. A measurement whose possible origin symbolizes a multicultural shared spiritual philosophy for a universal journey of humanity toward unity and higher consciousness. Again, the length of this measurement, this "spirit/light" cubit, is 27.5 inches (70 cm).

What is explored in detail and compelling evidence is offered for is a unit of measurement that unites both science and spiritual philosophy, Heaven and Earth, space and time. Such a statement should give one reason to pause, for it is tantamount; it did me, from the ancient civilizations, to the modern physics search for the "Theory of Everything" (TOE). In ancient times spirit and science were not approached as separate avenues of the understanding of our existence, the world, and the cosmos, but one united and synergized in an encompassing unity of the physical and metaphysical. The unit of measurement presented meets all these requirements.

Early on I realized it is important to ask; from where units of ancient measurements were derived. The answer is straightforward: body proportions. Think of the ubiquitous term used in the study of ancient metrology, the cubit. The cubit is a catchall term used for multiple units of measurements from multiple civilizations. Cubit is Latin for *elbow* and is a term used, in umbrella fashion, to categorize measurements approximately the length from elbow to fingertip, though these lengths range from anywhere around fifteen inches to twenty-eight inches. It is believed that almost all ancient linear measurements resulted from body proportions. This concept is exampled in Da Vinci's "The Vitruvian Man," an ink and paper drawing.

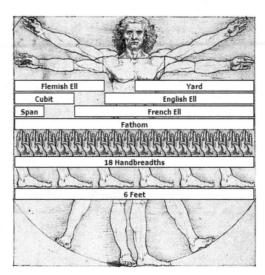

Fig. 3 DaVinci's Vitruvian Man with units of measure

This is DaVinci's homage to the Roman engineer and architect Vitruvius, who wrote in volume III of his work on art and architecture:

> *The design of Temples depends on symmetry... Hence no building can be said to be well designed which wants symmetry and proportion. In truth they are as necessary to the beauty of a building as to that of a well-formed human figure...*

> *If Nature, therefore, has made the human body so that the different members of it are measures of the whole, so the ancients have, with great propriety, determined that in all perfect works, each part should be some aliquot* (Author note: aliquot means a portion of the larger whole: I had to look it up) *part of the whole; and since they direct, that this be observed in all works, it must be most strictly attended to in temples of the gods...*[10]

Vitruvius is basically stating that sacred structures should use body proportions in their design and construction. Since the human body was divinely designed, these are the best proportions and measurements to design a sacred site, to create an axis mundi; a place where Heaven and Earth come together.

The linchpin at the heart of these sacred structures, these "cosmic engines" uniting Heaven and Earth, is the unit of measurement of 27.5 inches (70 cm). This unit of measurement seems to have its origins in ancient Egypt, yet there is compelling evidence that this unit of measurement or multiples of it was also used by other cultures on multiple continents, and so it presents itself as an ancient international unit of measurement commensurate to the modern international unit of measure, the meter.

Body proportions were codified into specific measures, such as the foot, the hand, or a yard. So from what proportions could 27.5 inches (70 cm) be codified from? Posited here is that this measurement comes from the length of the spine, with initial evidence from research in the medical textbook *Gray's Anatomy* (1918) that shows the average length of the human spine (male) is 27.9 inches (71 cm)—statistically valid to a measurement unit of 27.5 inches. In answer to the reader's thought, yes, there were many people of such height through ancient times. Some will debate that 27½ inches is close to the average step length and hence then where this measurement came from. What will be shared as we continue is clear and compelling evidence that this is not the case for multiple reasons. For now, let it suffice, using Vitruvius' recommendations of using body proportions since the body was divinely created, that a step length is not a body proportion. Further, there is an intrinsic elegance in representing the spine in sacred sites as the physical avenue of consciousness.

An Egyptian measuring rod of 27.5 inches (70 cm) was discovered at the pyramids of Lisht and is on display in the Metropolitan Museum of New York; Egyptologists link the pyramids of Lisht to the pyramids on the Giza Plateau through portions of the Giza complex being incorporated into Lisht to infuse them with the spiritual energy of the Giza complex. The Giza Plateau was dedicated to the Egyptian deity Osiris, whose symbol was the "djed" hieroglyph, meaning "Osiris's spine." The ancient Egyptians even had a specific ceremony rite of "Raising the Djed."

The recognition of the importance of the spine both physically and spiritually was not confined to the ancient philosophies of the Far East and Egypt. Its significance was acknowledged in the cultures of the New World also. Unambiguously this can be seen in the Hopi, Native American traditions. The Hopis are the descendants of the Ancient Puebloans, whose culture is considered to have spanned the Southwest area of what is now the United States: Colorado, Utah, Arizona, and New Mexico. One of the major beautiful and significant centers of the Ancient Puebloans is in Chaco Canyon located in northern New Mexico. Archeology research has confirmed that the Ancient Puebloans had trade and communications with Mesoamerica with the discovery of macaw feathers and remains of cacao traced back to Mexico.

The Hopi, descendants of the Ancient Puebloans, in their customs understood the importance of the spine both for their physical world and spiritual journey. This is recounted by Frank Waters:

The First People then understood the mystery of their parenthood. In their pristine wisdom they also understood their own structure and functions—the nature of man himself.

The living of man and the living body of earth were constructed in same way. Through each ran an axis, man's axis being the backbone, the vertebral column... Along this axis were several vibratory centers which echoed the primordial sound of life throughout the universe...[11]

...Palongawhoya (sacred twin), traveling throughout the earth, sounded out his call as he was bidden. All the vibratory centers along the earth's axis from pole to pole responded to his call; the whole earth trembled: the universe quivered in tune.[12]

It is abundantly evident the descendants of the Ancient Puebloan understood the symbolism of the human spine for uniting Heaven and Earth along with time and space.

The use and representation of the spine in the design construction and reason of sacred sites bespeaks of the elegance of purpose; to connect and commune with the Divine with the assistance of temple-structures designed for just such a consciousness-raising purpose. The Eastern traditions present this in a more straightforward fashion in attaining such a goal; through meditation and raising the Kundalini (coiled serpent), the spiritual cosmic energy, toward an awakening and spiritual transformation into higher consciousness. This occurs by raising this energy through the physical avenue from the base of the spine through the spinal canal to the brain. This concept and purpose of meditations is not the venue of the Eastern traditions alone; they were only the most straightforward in describing it. This purpose and the act of meditation is also in Western traditions, called by many appellations: contemplative prayer, Lectio Divina, and studying under the fig tree, to cite some examples. (See

Cistercian monks, St. Bernard, St. Malachi, St. John of the Cross, St. Teresa of Avila, mystic Judaism, Pierre Teilhard de Chardin, Thomas Keating, Aldous Huxley.) Meditation is being used today by businesses and corporations to assist their employees, only the practice has had stripped away any metaphysical connotations and is labeled "mindfulness," In any case its original purpose was to commune with the Divine and bring Heaven and Earth together within oneself.

What is written above has outlined the designs in constructing sacred sites to focus and assist people in connecting to the divine in consciousness, and it should be implicit such sites' designers and builders understood the totality of the journey, not just in spiritual consciousness, but also in physicality and in time and space. Many such ancient sites are known to be heavenly timepiece engines measuring seasons, solstices, equinoxes, moon phases, and more to great accuracy. Further, ancient time was measured by more than just seasons. Ancient Egypt is credited with creating the 24-hour day and the world's oldest hourly water clock was discovered in Egypt. These ancient architects and scientists had a clear grasp of time at many levels and linked Heaven and Earth through these sacred sites not only in consciousness but, in this manner, through physical space and time. What the following chapters will show is that ancient civilizations had an even more incredible grasp of time than previously thought. A grasp they incorporated into a much greater sense of cosmic unity.

Prior to continuing further in this journey, allow me to briefly provide some examples of this ancient unit of measurement (27½ inches/70 cm) from ancient cultures that will be detailed later on:

- **-Egyptian unit: Nebiu (NB) – 27.5 inches (70 cm)**... Possibly linked to **aakhu meh unit** and Great Pyramid; note the aakhu meh (transliterated as Spirit/Light cubit) is recorded in the ancient Egyptian hieroglyphs, but this research has yet to reveal any further information on it, other than its name.

- **-Paquime (Mogollon: Ancestral Puebloans) culture (Mexico-Arizona-New Mexico) – 27.5 inches (70 cm)** = 1 nebiu
- **-Mayan Cubit: Zapal – 55 inches (142 cm)...** Kukulkan Pyramid **= 2 nebiu**
- **-Stonehenge: megalithic rod:** (100 megalithic inches, 2.5 MY = 81.6 inches {207.3 cm}) = **3 nebiu 82.5 inches (210 cm)**
- **-Author research at Aztec Ruins National Monument great kiva. Great kivas at Chaco Canyon and Salmon Ruins kiva and rooms in New Mexico provide significant results using 55 inches (140 cm) = 2 nebiu or a zapal**

To return to Sir Petrie's method of examination of communication and sharing between civilizations: In *On Metrology and Geometry in Ancient Remains,* Petrie contends that measurement systems are an important proxy for divining the capacities of the ancient mind. The more complex the measurement system, he argued, the more complex the mind behind it. Additionally, Petrie established that measurement systems could be used as a method to evaluate connections between ancient cultures in a manner similar to the study of languages.[13]

Cultures sharing similar measurement systems likely had some form of contact. Should such a measurement be located in architectural remains, and appear in halves or doubles, then the probability that this measure reflects a real historical unit of measure increases.[14]

Sir Petrie in his later writings leaves no doubt of how important and substantial a unit of measurement shared by different cultures is in providing important evidence of an advanced mental capacity and either shared ancestors or strong trading links between such cultures.

> *Among the various tests of the mental capacity of man one of the most important, ranking in modern life on an equality of with language is the appreciation of quantity, or notions of measurement and geometry. … Thus the possession of the same unit of measurement by different people implies either that it belonged to their common ancestors or else that a very powerful commercial intercourse has existed between them.*[15]

What follows is a detailed examination of this evidence of a shared unit of measurement perhaps not only more important than language and more complex minds in that it also points toward a common effort to unite Heaven and Earth and time and space, with a unitive spiritual philosophy aimed toward higher consciousness among cultures strewn far and wide and the premise that ancient civilizations on at least three different continents had communication with each other farther back in time than is presently considered possible—all of this "hidden" in plain sight.

Scholars Iain Morley and Colin Renfrew understood and expounded on the universal seeking to unite and interact with the physical and the spiritual, the material and immaterial.

> *From the stones of Stonehenge to the alignments and calendars of Mesoamerica, measurement stands at the dawn of cosmology. The term "cosmology" is used here not just in the sense of explanation of the celestial, but in the sense of conception of the universe—the set of beliefs about the world, material and immaterial, and the rules through which interaction can occur.*[16]

I consider this measure, called by some Egyptologists nb or nebiu, is also identified in ancient Egyptian texts as the "aakhu meh," which translates as spirit or light measure and that it was communicated to multiple cultures. For the purposes here, I will call it the spirit/light cubit.

I hope this study does for you what it did for me in my searching, delving out into the physical world of sacred sites and in to the spiritual world of higher consciousness, bringing a peace and hope and flow into my life. It is like being cured of color blindness and beginning to see the vibrant hues of the universe.

Wonders of a lifetime/ Right there before your eyes/ Searching with this life of ours/ You gotta make the journey out and in/ Out and in, out and in

"Out and In" by The Moody Blues Album: *To Our Children's Children's Children* 1969

3

Humanity's urge to measure and incorporate space and time in their consciousness — squaring the circle

Time lives our lives with us; Walks side by side with us
Time is so far from us; but time is among us
Time is ahead of us; above and below us
Standing beside us; And looking down on us.
　　　　　　　　　"Time Song" The Kinks 1968 (2018)

Fundamentally, since our early evolving consciousness, human beings seem prone to measuring and identifying ourselves and our place from our daily existence to the cosmic scales in the pattern of the universe. Science considers our physical existence generally in a four-dimensional arena consisting of three spatial dimensions and one of time. Even time is often related to with spatial concepts such as being backward (past), forward (future), and present (here) for measuring temporal distances.

...it is clear that the origin of number, like the origin of language, is closely connected with the way in which our minds work in time...Our idea of time is thus closely linked with the fact that our process of thinking consists of linear sequence of discrete acts of attention.[17]

Professor Whitrow points out, our consciousness early on attempted to adapt time measurement to the linear measurements of the physical world around us. In daily life the sun went in a line across the sky, the shadows grew longer in line as the day passed, and the moon and stars followed suit in their own tracks. Events, occurrences, and acts followed one after the other. Yet as such observances continued, time would reveal to us not only linear concepts but non-linear cyclical events also. The sun would rise and set and come back around, the moon would rise and set and move through phases and come back around, seasons would come and go and return, and stars would do the same, ad infinitum. Perhaps these cyclical events in the heavens beyond our reach versus the linear spatial realities of Earth literally below our feet was the progenitor of a consciousness that incorporated spiritual and metaphysical aspects that went beyond space and time, a perception that was greater than that. Imaginably resolving this apparent dichotomy expanded our consciousness in order to square the circle and bring Heaven and Earth together.

Indications toward this expanded consciousness can be seen in the cave artwork of prehistory. At sites in France, Spain, and Indonesia, incredibly prolific and beautiful cave art has been discovered and dated from periods ranging from 37,000 to 15,000 years ago. Generally this art has been interpreted as depicting possibly past hunting successes or rituals to help in ensuring future hunting success. Though there is a clear challenge of interpreting the thought behind such prehistoric art, it is not extreme to consider that these artists understood such concepts as past and future being separate from the present, along with metaphysical, shamanistic types of concepts. It is well within the realm of possibility that these ancient prehistoric people's consciousness was evolved to the point of grasping a thought process of time and space along with more ethereal concepts and symbolism.

(Author's note: With a 27.5-inch spine, a person would be approximately five feet, nine inches tall, and yes, there are multiple examples of ancients being that tall. Researchers studying these ancient cave paintings have determined that at least one of these extraordinary artists was six feet tall.)

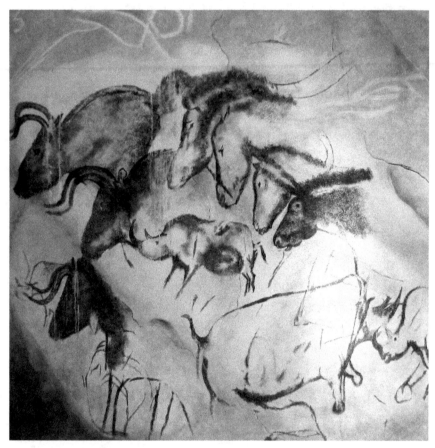

Fig. 4 Replica of the painting from the Chauvet cave, in the Anthropos museum, Brno. (30,000 – 28,000 BC)

Further research into the consciousness creating such cave art and its purposes reinforces such interpretation of higher and more complex consciousness being at work. This is exampled by Dr. Michael Rappenglueck, a professor at Munich University and an archeoastronomer, which is one who studies and investigates astronomical knowledge of prehistoric cultures. In his study of the cave art at Lascaux, France, he interprets some of the cave art to be symbolic depictions of the stars in the heavens, specifically what is now known as the Summer Triangle.

According to Dr. Rappenglueck, these outlines form a map of the sky with the eyes of the bull, birdman and bird representing the three prominent stars Vega, Deneb and Altair. Together, these stars are popularly known as the Summer Triangle and are among the brightest objects that can be picked out high overhead during the middle months of the northern summer. Around 17,000 years ago, this region of sky would never have set below the horizon and would have been especially prominent at the start of spring. "It is a map of the prehistoric cosmos," Dr. Rappenglueck told BBC News Online. "It was their sky, full of animals and spirit guides."

The archaeologists who have looked at Dr. Rappengleuck's conclusions have so far agreed that they are reasonable and that he may have uncovered the earliest evidence of humanity's interest in the stars.[18]

This cave art and its interpretations is just one illustration of early human beings having a complex and evolved consciousness— a consciousness that evidences a strong grasp of time and space, astronomy and heavenly events, along with symbolism, abstract thought, and a shamanism of metaphysical concepts beyond four dimensions. It is important to remember these are just a few discovered examples of an evolving human consciousness occurring 30,000 years ago (i.e., burials, jewelry) if not earlier.

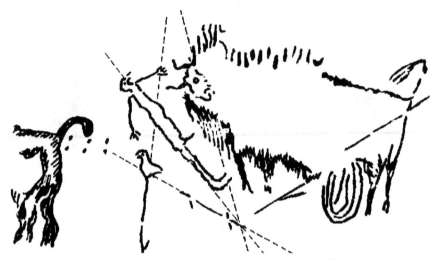

Fig. 5 Rendition of Lascaux cave art interpreted as the Summer Triangle stars (17,000 BC)

A continuous timeline and examples of evolving human consciousness could be shown, but the evidence is already clear, so allow me to fast-forward about 25,000 years to Egypt approximately 5,000 years ago to a period presently considered the end of prehistoric times and the beginning of written history of "recorded time." Ancient Egypt is considered one of the earliest "cradles" of civilizations, long believed to be where the burgeoning of some of the "first" advanced civilizations began. It is in Egypt that I have found, so far, the earliest origination of the 27½-inch unit of measurement. In a relatively short period of time (just hundreds of years), the Egyptians advanced in a myriad of avenues such as writing, science, technology, engineering, and architecture to construct what is considered one of the ancient wonders of the world, one that remains a wonder in the modern world today: the Great Pyramid.

In my personal experience the Great Pyramid—its design, its construction, its purpose, its ability just to awe and make anyone wonder that sees it, stand by it or enters it; that is where my journey finds its touchstone, its compass. The next chapter will go into more detail and specifics concerning this amazing structure along with other ancient sacred edifices around the world. In this chapter the focus will be the unit of measurement coming out of Egypt, possibly used in the Great Pyramid, and a clear case for a 27½-inch unit in its design. It will show its relation as a model for an ancient integration of this unit of measurement and its relationships to the pendulum and time. I think you will see just how fitting the Arab proverb truly is: "Man fears time, but time fears the pyramids."

As I first explored ancient Egypt, once I could pull myself from the Great Pyramid, I was drawn by the Egyptians' incredible achievements, Egypt's mysteries and enigmas, and all the esoteric tales going back into its sands of time. From biblical accounts to the ancient scholars from other civilizations, wise men wrote of their pilgrimages to Egypt to draw and gain knowledge from the Egyptian teachers and priests. Modern scholars and researchers are still sifting through that ongoing mystery called ancient Egypt. An interesting point is that the earliest historians of Egypt only go back to approximately 500 BC, 2,500 years *after* what is considered the beginning of their dynastic periods; 500 BC is at the end of Egyptian native rule and the beginning of a 2,500-year march of foreign rule from Persians, Greeks, Romans, and Muslims. Amidst this fading descent of native rule for 1,500 years, approximately from 300 AD until the early 1800s, until the translation of the Rosetta Stone, the ability to read Egyptian hieroglyphics had been lost and still hasn't fully returned. One has to consider how much of the original ancient Egyptian knowledge and philosophy has been lost through all this. Think of the children's game of "telephone," passing a message on from person to person through multiple people and how often the message has changed from its beginning communication to the end. Now imagine this occurring over thousands of years and what would be left of the original message. Even the shifting sands of the Sahara seem to have collaborated to conceal this information as Egyptologists consider that at least two-thirds of ancient Egypt is still buried beneath.

In the beginnings of my journey, in reading and exploring all of this, one source I read mentioned that a unit of measurement used in the construction of temples and pyramids of Egypt was 27½ inches. I thought this would be easy to research. Sigh, silly rabbit, and down the "rabbit hole" I went.

I thought it would be straightforward to research an Egyptian unit of measurement of 27.5. I was wrong. In almost all searches for Egyptian cubits and measurements for pyramid and temple construction, two results will be returned: a "little" cubit (netches meh) of 17.5 inches and a "royal" cubit (nesu meh) of 20.6 inches, with the royal cubit considered to have been the measurement used in the Great Pyramid and others. If I hadn't already read about the possible existence of another "cubit" of 27.5 inches, I probably would have accepted the mainstream opinion and truly would have been none the wiser because of it.

With a strong-willed (stubborn) belief in this alternative measurement of 27.5 inches, I began my research, rather than digging in the sands of Egypt, it was through the dust of archives and databases into the deep recesses of Egyptology. I finally had my "eureka" moment, but rather than finding "wonderful things," I discovered a "wonderful measurement": a 27½-inch Egyptian cubit. Not only did I find evidence of it in Egyptology, but it led me to an actual physical specimen of the measuring rod in a major museum. This 27½-inch Egyptian measuring rod can be found in the Metropolitan Museum of New York. (See Chapter 4.)

The bottom line is that this measurement can be documented not only back to Egypt, but it also can be found evidenced in multiple cultures across the globe as was presented in Chapter 2. Perhaps you are asking yourself, as I did, how is this possible? The remainder of this chapter will begin to answer to that question, first, with the focus on how this Egyptian measurement appears to be an international unit of measurement akin to today's meter, along with startling synchronicities with them both and the pendulum.

As I uncovered the use of this same unit of measurement in diverse cultures—one that appeared to be a measurement shared across cultures and continents—as I had previously questioned where ancients' measures came from, I now asked where today's modern meter originated from. I tend to always ask why in an effort to reach back into more empirical answers. It made me smile and think back thankfully on the patience of my parents dealing with a child who was continually asking why and their kindness in putting up with such a quizzical child.

While the discovery and research of this ancient Egyptian measurement had turned into an incredible saga of adventure with international ramifications for much earlier global communications between cultures and a shared spiritual consciousness, the tale of the creation and acceptance of the meter is a story that reads like a plot from a John le Carré novel. The meter was 200 years in the making, leading to a 700-mile journey of seven years that was steeped in international politics, war, civil war, mathematicians, astronomers, intrigue, and finally hidden errors in its creation.

To really appreciate how this ancient measurement could have been the world's first international unit of measurement and how it was created in a similar fashion as the meter, a brief history of the formation of the meter is needed. It started in the late 1600s with several scientists of the time who began calling for the need of international units of measure primarily to standardize scientific inquiry and international trade. The initial and leading suggestion for this unit of measurement was to use the length of a pendulum whose to-and-fro swing (period) equaled two seconds—one second for the swing in one direction and one second for the return swing, known as a "seconds" pendulum. From the 1600s to 1930, pendulum time pieces were considered the most accurate in the world. The length suggested for the meter was 39¼ inches (997 mm), the length of the pendulum that when swung at less than a 20-degree arc became a seconds pendulum. It would take over another one hundred years before the suggestion for an international standard would really begin to move forward.

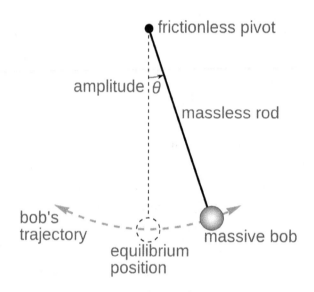

Fig. 6 Pendulum

At that later time, while the French revolution was occurring along with the birth pangs of their "age of enlightenment," the push for an international unit of measure began again. This time the discussion was whether to use the pendulum length as a unit of measure or a fraction of the length of the distance along a meridian (longitude) from the equator through Paris to the North Pole. This original meridian through Paris suggestion was changed for international support and acceptance. The pendulum was considered a good choice in its portability and the ease of recreating it all over the world. Think of the metronome used in keeping time for music or a grandfather clock; they use the same principle and do not hinge on any claim to a particular country due to a geographic line. The suggestion had international backing from France, England, Germany, along with the new country of the United States of America with Thomas Jefferson leading the support for its use. Thomas Jefferson saw this length derived from the pendulum having the potential to *become a line of union with the rest of the world.*[19] I find Thomas Jefferson's quote eerily echoing the purpose of this ancient spirit of light cubit's goal. The support for the pendulum also had it detractors who argued that due to gravitational discrepancies in different places on the earth, the pendulum would not be accurate enough.

(Author note: The discrepancies argued were of a difference of a one-half of one percent gravitational effect measured at the pole versus the equator, with lesser differences occurring between the pole and equator.)

The supporters of determining the meter from and as a fraction of the earth's distance from the equator to North Pole rather than using a seconds pendulum was that; "only a measure taken from nature could be said to transcend the interests of any single nation, thereby commanding global assent and hastening the day when the world's people would engage in peaceable commerce and the exchange of information without encumbrance.[20]

Purportedly, due to the argued gravitational discrepancies, the French National Academy of Sciences chose to use one ten-millionth of the distance from the equator to the North Pole. This would be the meter and would be determined by an expedition measuring the distance between Dunkerque (Dunkirk), France, through Paris to Barcelona, Spain. Approximately 90 percent of this distance was in France, using the Paris meridian. This decision from the French caused the collapse of all the international support for a shared unit of measurement, it being seen as a violation of any single country laying claim to an "international" measurement. The French continued alone and funded a seven-year-plus expedition that was commissioned to get this equator-to-pole measurement using incredibly accurate astronomical observations and mathematics over this chosen distance, primarily through France, and extrapolating from it the distance from the equator to the North Pole. Curiously the resulting derived length was only three millimeters different than the proposed length derived from the pendulum, and that does not take into account the errors, though miniscule, in the calculation hidden away for 200 years before being discovered by author and researcher Ken Adler.

The meter's fascinating history continues down a long, convoluted journey, and its length is now evolved to be the distance light travels in a vacuum in 1/299,792,459 of a second. What is so fascinating in the birth of the meter and its evolution are the synchronistic parallels it has with this Egyptian measurement thousands of years older, with relationships from the pendulum to a measure of light that will be shown.

I posit that the Egyptian measurement of 27½ inches was used in a similar manner as today's meter for an international unit of measurement in what is considered prehistoric time. Besides the structures themselves that provided standing evidence of its use in Egypt, Great Britain, Mexico, and the Southwest of the United States, there is further evidence provided in the history of these ancient cultures and in the history of the modern meter.

As described before, the way to determine the length of the meter boiled down to two final choices: using a tiny fraction of the measurement of the earth's surface derived from the distance from the equator to the North Pole, or using the length of a seconds pendulum. Each recommendation had its own drawbacks, but I think about which one would be the easiest to share and re-create among other varying cultures and could be passed on even if a civilization fails. I believe the answer is clear, it would be the pendulum. The next question is what about their suggested length of 39 and ¼ inches for the "seconds' pendulum", that still doesn't fit a unit of 27 ½ inches! That is correct, for a pendulum that is swung at less than 20 degrees.

So I researched what length a pendulum should be if swung at a different degree angle and still create a seconds pendulum. Incredibly what I discovered was that starting the swing of a pendulum at 90 degrees (at the horizontal), you will get this all-important two-second pendulum at 27.5 inches! Incredible!

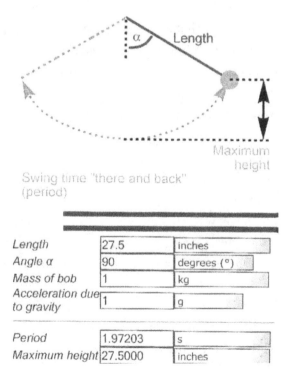

Fig. 7 Pendulum software program determining length

Length	27.5	inches
Angle α	90	degrees (°)
Mass of bob	1	kg
Acceleration due to gravity	1	g
Period	1.97203	s
Maximum height	27.5000	inches

I would like to thank all the universities and their "edu." websites and other sites that allow the public access to their resources. I am forever grateful for how important and helpful they were in my research. Using mathematic software programs to test this theory gave a correct pendulum length of 27½ inches if the pendulum swing was started at a 90-degree angle. This provides a simple and elegant solution of placing the apparatus in the ground, then bringing the pendulum up to the horizontal (90 degrees) and letting it swing. Just as in the suggestion of creating the international unit of the meter, this also creates such a measuring unit.

Think how simply ideal this is. A 27½-inch pendulum, swung at the horizontal, creates a seconds pendulum, not at 35 degrees, not at 50 degrees or at 70 degrees, but at 90 degrees, the horizontal. The horizontal equates it to the horizon—an important concept and aspect of ancient civilizations. The horizon or horizontal is easy to find, easy to remember, easy to recreate.

Ancient civilizations were aware of pendulum action; the known use of plumb-bob type devices in their building construction makes this evident. The horizon was also important to these civilizations, with much of their astronomy using the horizon to note heavenly movement. One of the ancient names of the Great Pyramid is "the horizon of Khufu." The ancient Egyptians also used the hieroglyph "akhet" to represent the horizon and moreover called it "Place of Glorification."[21] It is clear the horizon had significance to them, symbolic of where Heaven and Earth come together, and such significance would translate well in the incorporation of a time and space measuring pendulum swung at the horizontal.

This is even more appropriate when one examines the survey equipment that may have been used not only to build such structures, but also to measure and transit the stars above. In the case of the Egyptians, this was a plumb-bob type of device called a "merkhet," which is translated, appropriately, as an "instrument of knowing" or

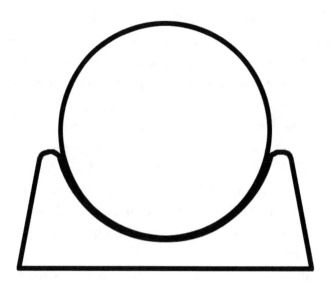

Fig. 8 Akhet

an "instrument of time keeping." It was known to be used for time keeping by measuring stars, for finding north, and for building construction. This was very similar to a surveyor's tripod and plumb bob and could have also functioned as a pendulum.

To the question of whether the Egyptians recognized and used the pendulum as a timepiece, there is a very telling and important quote from Sir Flinders Petrie, a recognized icon of modern Egyptology.

> *Whether the Egyptian treated the well-known plumb-line as a pendulum is not indicated by any remains, though the plumb-line was commonly in use from very early times. But the notable fact is that 29·157 ins. (the diagonal of the 20·62 ins. cubit), which was the basis of all land measure, is the length which would swing 100,000 times in 24 hours, exactly true at Memphis latitude. This is so remarkable that it suggests that it may have been derived from that observed length, and the source entirely forgotten after the scientific age of the pyramid builders.*[22]

Sir Flinders Petrie, not only an icon of modern Egyptology but also considered by many as the father of modern Egyptology and the use of the scientific method in Egyptology, posits that the Egyptians were aware and used the pendulum as a timepiece and that this knowledge was lost, dare I say, in the sands of time. The theory proposed here for such a pendulum timepiece with a length of 27½ inches could barely have a better pedigree.

That being written, it has been noted Petrie also suggested a length of the pendulum of 29.157 inches, the diagonal of a square made by using the royal cubit length of 20.62 inches. Further, this pendulum would swing 100,000 times (Author note: 50,000 periods, a swing out and back) in 24 hours.

The validity of the theory that the Egyptians used a pendulum as a 24-hour timepiece has been confirmed here by Sir Petrie. Presented next will be the evidence and rationale that they used a pendulum of 86,400 swings or 43,200 periods, rather than a pendulum of 100,000 swings or 50,000 periods. Any length pendulum with its arc swing will give some count of swings and periods for 24 hours.

The key to this discussion is the length and arc swing, primary factors in determining how many seconds and fractions of seconds its swing and period are. This can be seen in Sir Petrie's example. His length (29.157 in.) swung from about 20 degrees would result in 100,000 swings or 50,000 periods. Sir Petrie connects it to the royal cubit and the length of the diagonal of a royal cubit square. Advanced here is that it is not a diagonal determination; rather it is created from the same length of the Egyptian cubit, the nebiu (27½ inches, aka spirit/light cubit). As noted earlier Egyptologists generally believe the royal cubit was used in the construction of the Great Pyramid (aka the *horizon* of Khufu). As shown earlier there is an elegance and cultural significance with a pendulum swing started from a horizon(tal) position and for the use of such a length that measures time and space. In reference to units of measurements used in the building of the Great Pyramid, the next chapter will present evidence that the 27½-inch unit was used in the Great Pyramid with even Sir Petrie allowing for this possibility.

Fascinatingly there is more evidence for this unit provided by the dimensions of the Great Pyramid of Egypt itself. Several researchers have noted that the Great Pyramid appears to be a geodesic model of the earth, a 1/43,200 model. Different dimensions of the Great Pyramid, such as its perimeter from its corners and the perimeter from its socket holes, can be used to create a very accurate 1/43200 Earth model at the equator latitude circumference and of the longitude circumference through the poles of the earth. Dr. Robert Schoch points out this equator circumference scale for the model and believes it was chosen due to it representing the length of one-half day being 12 hours represented in seconds (43,200 seconds).[23]

I wholeheartedly agree with Dr. Schoch and the other researchers' conclusion of the Great Pyramid being a 1/43200 scale model of Earth's dimensions and that its representation is linked to time, specifically the amount of time related to a solar (24-hour) day. If that is the case, would the builders stop there? Would not this scale and purpose and symbolism of uniting time and space, Heaven and Earth, be incorporated further into the dimensions of the Great Pyramid? I say yes. I have merely continued the apparent purposes of the ancient builders, positing they used a measure of 27.5 inches. Remember, the beauty here is that the seconds pendulum—a pendulum with a two-second period—is equal to 1/43200 of a mean solar day, repeating this period 43,200 times in 24 hours as the 27½-inch horizon swung pendulum suggested here, matching the suggested geodesic model of the Great Pyramid, not only as a half day though, but representing the journey of the sun for a whole day. After all, the ancient Egyptians are credited with giving us the 24-hour day. Using this spirit/light measure gives significant results in its units of measurement not only for the Great Pyramid, but as a model of the Earth and keeper of time.

Grant me your patience to reiterate on the use of this measurement for space. It has also been shown to have been symbolically important to the Egyptian spiritual philosophy by representing the spine and also functions beautifully as a timekeeping pendulum from the horizon(tal). This all draws together to an integrated, elegant conclusion for this scale model of Earth and keeper of time and in so doing unites time and space, further in representing the spine (the djed) and the ritual of "raising the djed," which unites Heaven and Earth. Imagine the ramifications for us today that an ancient civilization, thousands upon thousands of years prior, to us had such evolved sciences and spirituality. Imagine what still is lost, yet to be found, what is still to be learned—it's stunning.

The ancient Egyptian consciousness gestalt truly represents an encompassing philosophy whose whole is greater than its parts, but the integrated parts are fascinating in themselves. I include some such final synchronistic parts:

- The formula to determine the time period of a simple pendulum is $T = 2\pi\sqrt{l/g}$. The perimeter of the Great Pyramid over its height is 44/7 or 2 pi. The Great Pyramid and the pendulum seem to share the same mathematical constant of 2 pi, which is also used to determine the circumference of a circle.
- The history of the meter has been briefly presented. Today, for better accuracy, the length of the meter is determined by how far light travels in a vacuum in 1/299,792,456 fraction of a second. Talk about retro-engineering a measurement to fit! The pyramids of Egypt along with their purpose and construction were always related to light. Obelisks (aka tekhenu) represented frozen rays of light; the purpose of pyramids

was to unite the soul's ka and ba with the body to create an aakhu, a radiant being of light. In the hieroglyphs there is a mysterious unit of measurement called an aakhu meh (spirit/light measure). There was no retrofitting by the Egyptians for their light measurement. This ancient measurement was related to the concept of light both in physicality and spirituality from the beginning.

- Finally, it was just noted how the mathematical constant pi, specifically 2 pi, was related to this ancient measure, the pendulum, the dimensions of the Great pyramid, and determining the circumference of a circle. Further, I had referenced Da Vinci's artwork "The Vitruvian Man" earlier in reference to body proportions and measurements. The Vitruvian Man is also known for "Squaring the Circle." In the mathematics of geometry, this was an impossible feat to do, but Da Vinci, in the Vitruvian Man, accomplished this for human beings. This can be seen in his art with the human body bracketed by both the square and circle. The mathematical impossibility was already known; this depiction was symbolic of bringing Heaven and Earth together with oneself, the square symbolizing the earth and the circle symbolizing Heaven. The Vitruvian Man is an example of a human body being an axis mundi, a point where Heaven and Earth come together. Similarly Egyptologists describe pyramids as representations of where Heaven and Earth unite. This presents even more of a reason to use this suggested spirit/light measure.

In the many of descriptions written here, the term *elegance* or *elegant* is used, not so much for describing the attractive and graceful accomplishments of the ancient civilizations, though that is evident, but more in the mathematical sense. In mathematics and elegant equations is a formula that is straightforward, precise, succinct, and powerful. Think of Einstein's famous formula, $e=mc^2$, an equation that has incredible universe- and earth-shaking use and implications, an equation that basically created a new paradigm in physics, yet consists of just five mathematical symbols. That is mathematical elegance. In its own way I perceive that simple unit of measurement, 27.5 inches, in the same fashion. A simple length, yet symbolic of the spine, symbolic of higher consciousness, functioning both as a measure of distance and of time, all this at once—that is elegant, powerful, and far-reaching.

Quoted earlier, the French Academy describes the meter creating a measure that "transcends the interest of any single nation, thereby commanding global assent and hastening the day when the world's people would engage in peaceable commerce and the exchange of information without encumbrance."

I describe the ancient measurement postulated and described here, in its elegance, representing the spine and unifying not only time and space but also Heaven and Earth; a measure that not only transcends any culture's physical sites but one that **is transcendental for human consciousness!**

Humanity's urge to measure... 41

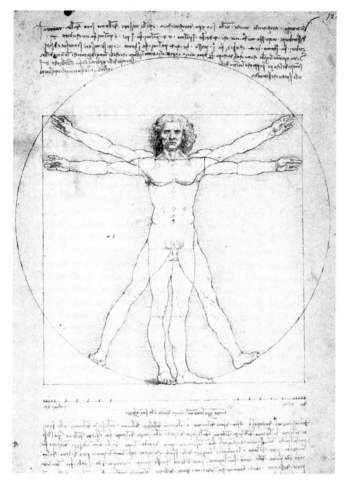

Fig. 9 Leonardo DaVinci's Vitruvian Man

It's a pendulum, it all comes back around
It's a pendulum, it's a pendulum
Life's a pendulum, it all comes back around
It's a pendulum, it all comes back, it all comes back
"Pendulum," Kate Perry 2017

4

Sacred structures connecting Heaven and Earth; the measurement results

a. Egypt
b. The Maya
c. North America's ancient Puebloans
d. Stonehenge

*Y*ou have already read from the scholars the importance of the study of ancient measurements in understanding that shared measurements not only provide evidence of communications between societies but also the ancient cultures' thought processes. That such measurement was an application of their thought processes to unite the physical and spiritual worlds. I ask the reader to pause for a moment and ponder the idea that ancient cultures created and shared a specific measurement that was transcendental in its purpose to address both their physical world and spiritual world; in one measure they united time and space, Heaven and Earth. This ability goes directly to the purpose of religion (putting aside the discussion of religion versus spirituality). One etymological interpretation of the word "religion" is to unite or create a bond between humans and gods, just as these ancient cultures are evidenced of doing. The elegance and immensity of this ancient consciousness is awe-inspiring and humbling.

At this point this chapter may seem almost anticlimactic, but before moving into the infinite transcendental symbolism of this measurement, first its finite physical application for constructing sacred sites, sites representing an axis mundi for each culture, needs to be clearly presented. As these ancients first built their "cosmic engines" in earth to facilitate reaching higher realms of consciousness, connecting the Heavens to Earth, so first must I complete the physical evidence of the use of this measurement. In writing, I share that the decades of seeking, research, and discovery in this journey is a gestalt-like experience, and putting it in as linear a format as possible has been a challenge. Some of this will be redundant, and some will cross over; it is necessary. This chapter will be subdivided into sections on Egypt, the Maya, Puebloan cultures, and Stonehenge.

Each section will provide the archeological academic findings for each of these cultures; the findings will demonstrate the substantiation that such a transcending unit of measurement existed and was used by each of these cultures in the design of their sacred sites. Also it will be established that each of these cultures understood this measure was derived from the length of the human spine, and it implicitly was symbolic of the channel of spiritual energy, known in the Far Eastern spiritual traditions as the Kundalini, the serpent energy. I think by the end of the chapter you will find, as I have on this journey, that it all has been hidden in plain sight.

A: Egypt: Crucible of Civilizations

I call Egypt a crucible of civilizations rather than a cradle. For me, in this context, a crucible is a place of concentrated influential forces that inspire and impact the development of many other cultures. In my journey Egypt meets that definition and, as of yet, the earliest I can document this measurement, is its genesis.

At present Egyptology recognizes the general use of two different Egyptian cubits as units of linear measurement in the civilization of ancient Egypt. They are the "royal cubit" (nesu meh) of 20.61 inches (52.35 cm) and the "little cubit" (netches meh) of 17.5 inches (44.45 cm). It is generally accepted that the royal cubit was used in the construction of temples and monuments. An example of this is the Great Pyramid of Giza, which when measured by this system has a height of 280 royal cubits and a side width of 440 royal cubits.

What is being presented here is the evidence for the possibility that a different unit of measurement, a different length cubit, may have been used in the construction of pyramids. It should be noted that the word "cubit" is a Latin word meaning elbow and has been used as a catchall term for many ancient civilizations' measurements, often having more than one measurement deemed a cubit; it is not the term the individual cultures used.

The different unit of measurement put forth is approximately 27.5 inches (69.85 or 70 cm) this consideration is based upon the purpose of pyramids themselves and an identification of a "third" Egyptian cubit, considered a non-standard cubit. Evidence of this third cubit consists of physical discoveries of it and its mention, in different forms, in the hieroglyphs. How this third cubit is named in the hieroglyphs also ties into the purpose of the pyramids and an early deity of ancient Egypt whose name and purpose also link to the name of this third cubit and the pyramids.

I had already noted reading of an Egyptian measurement of 27.5 inches (70 cm) yet was initially finding no evidence for it. My breakthrough was found in the writings of Dr. Livio Stecchini, a Harvard PhD and an expert in ancient measurements. In his treatise *A History of Measures*, Part II, Chapter 3, he writes:

> *More recently one has come to realize that a number of documents indicate the use of a unit equal to 1 1/3 royal cubit. In my opinion this mysterious unit is a cubit of two hybrid feet, that is, 37 1/3 natural basic fingers, 700 mm.* (Author's note: 700 mm = 70 cm) *The name of this unit is nb, nebiu, which means "carrying indicates that the original unit of length was the carrying yoke; the term for cubit in Semitic languages and in Greek(-) means the arm of the carrying yoke, that is, the half of it. On Egyptian cubit rules, the position of the hybrid foot is indicated by the sign of the forearm rmn; the term means "cubit," but it corresponds to the idea of "to carry" and it also means "half," indicating that essentially it signifies the half of the carrying yoke.*[24]

Now I had my first firm academic evidence for this Egyptian measurement and a name for it, providing a thread for me to follow. But Dr. Stecchini's writing provided even more; it included the statement that an actual physical measuring rod artifact of this was in existence.

> *My explanation of the unit nebiu is supported by a neglected specimen of the Metropolitan Museum of New York. This object is listed in the catalog as a cubit of 27½ American inches (698.6 mm). It is a double hybrid foot (rmn) or a "carrying yoke" of 700 mm. It consists of a simple round rod of plain wood divided by lines cut with a saw into 7 parts; the seventh at the middle is further divided into two parts, so that the rod is divided at the center in two halves of 3½ sevenths.*[25]

Fortune was smiling on me, for not only now did I have academic evidence, but there was a physical specimen of it reported also. My fortune both in the physical and spiritual continued. My youngest son was living in New York City at the time, and at my request to check the Metropolitan Museum Art of New York for this measuring rod, he found it! I offer many thanks to my son for humoring his father. The MMA is a great museum with a large multilevel Egyptian section and exhibition, and I can attest to this because I needed to go and see it with my own eyes. I could have easily missed this measuring rod tucked away in a dusty corner of a dead-end corridor. I thank you again, Corwin.

Fig. 10 Egyptian 27.5-inch measuring rod at the MMA Gallery 109

This measuring rod was discovered during the excavations in the Lisht area of Egypt, and is related to the pyramids of Amenemhet I and Senusret I of the Middle Kingdom, more particularly to the north side of el-Lisht, the Amenemhet I Pyramid. This is an interesting note in that Amenemhat I, the first ruler of the 12th Dynasty, was trying to revive the Old Kingdom style of pyramid complexes (i.e., the pyramids of Giza Plateau) and their spiritual purpose as described by Dr. Mark Lehner:

> *Picking up the pieces to resurrect the pyramid age: Amenemhut I incorporated fragments of Old Kingdom tombs and pyramid complexes in his own pyramid."*[26]

> *"Amenemhut I returned to the approximate size and form of the late Old Kingdom pyramid complex...Perhaps the most remarkable feature is the fact that it included fragments of relief-decorated blocks from Old Kingdom monuments—many from pyramid causeways and temples, including Khufu's....We can only conclude that they were picked up at Saqqar and Giza and* **brought to Lisht to be incorporated into the pyramid for their spiritual efficacy.***"*[27] (Author's emphasis.)

Dr. Lehner concludes that fragments from Old Kingdom pyramids, including those of the Giza Plateau, specifically the Great Pyramid (Khufu's), translate and infuse in the Lisht pyramids the spiritual energy and provides an important link for this measurement to the Great Pyramid.

Now with these initial threads, I was able to find more documentation of this spinal-based (if I may call it so) measurement. There is documentation of its use at Old and Middle Kingdom sites. I share an excerpt of one such document below.

> *...the non-standard measuring rods of 65-70 cm, which were discovered by Petrie at Kahun (Author note: Kahun was a temporary living site for the workers building the pyramid at Al-Lahun, and was built under the reign of King Senusret) and at Deshasha (Author note: site of Old Kingdom rock-cut tombs), and another similar rod from Lisht. Now it has long been considered that a measurement of between 65 and 77 cm could be equated with a unit known as the nbi, which is seldom*

mentioned in the Egyptian literature and was generally used to record the amount of work carried out in the cutting of dykes or the excavation of tombs. Gardiner conjectured that the nbi might be equal to 1 1/4 or 1 1/3 cubit,[3] and thus gave a name for the measurement of 1 1/4 cubit in the tomb of Tausret.

At this point, however, we have to contend with two recently published alternative explanations for the nbi measure, **both agreeing that the length should be 1 1/3 cubit or 70 cm** (Author's emphasis) *in opposition to Elke Roik's proposed value of 65 cm, but differing in their interpretation as to the use of this measure. Following his survey of a number of Old and New Kingdom rock tombs, Naguib Victor maintains that the nbi had an architectural significance;[4] while Claire Simon believes that it was connected with the canon of proportion, and was used to determine the size of the grid squares in which the human figure was inserted.[5] The existence of a linear measure of about 1 1/3 cubit is in any case proven for the Middle Kingdom by the three rods from Kahun and Lisht, which vary in length between about 67 cm and 70 cm. These rods are clearly too long to answer to Roik's measure of 65 cm; and since each is divided into seven units, they cannot supply the required dyadic divisions.[28]*

Besides citing the multiple sites where this measurement has been found and also confirming Dr. Stecchini's research, it is the discussion of this measurement's proposed use that caught my excited attention. Researchers in the article consider that the measurement is more accurately codified at 70 cm (27.5 inches) and its use is considered separately either in the architecture itself or being linked to outlining human proportions for the depiction of the human figure. As presented in this book, this measurement based on the spine was not an either/or proposition of architecture or art body proportions but both and carried with it the greater symbolic spiritual energy.

Further, the use of such a proposed unit of measurement, at least for the Great Pyramid, is given support by the research of Egyptologist Sir Flinders Petrie.

> *The predominance of mason's* ***measures in the Great Pyramid has suggested that a variety of measures were in use, some of which do not seem to be an even number of digits.***[29] *(Author's emphasis.) So there is evidence for independent standards which are not formed from digits. This may seem unsatisfactory to anyone expecting a cut-and-dried result; but the subject is new, and it can only grow by fresh facts which agree with what is already explored.*[30]

The possibility and plausibility of this unit of measurement in the Great Pyramid should not be ignored.

Even its name, nhb, nb, nbi, demonstrates links to the Great Pyramid. As earlier stated the Egyptian root word nb translates as to unite or yoke together. The ancient Egyptians recognized a predynastic deity whose purpose was to perform similarly this function. The deity's name was Nehebu-Kau (also spelled Nehebkau and Neheb ka). Nehebu-Kau, "He Who Unites the kas," was a benevolent serpent god who the Egyptians believed was one of the original primeval gods. This ancient deity is generally depicted in various anthropomorphic forms, but he was also depicted as a serpent and on occasion as a serpent with a head at each end of its body, clearly symbolic. This will become more significant later on in the discussion of the Maya culture. He united the ka (life force) and the ba (individual soul), and in this reunion they become the akh (radiant light), becoming a celestial, immortal being.

The purpose of this benevolent serpent deity of yoking together, uniting the soul, not only exhibits a link to this named measurement, but Nehebu-Kau's purpose directly reflects the design function of pyramids, including the Great Pyramid. Dr. Mark Lehner, an Egyptologist and considered one of the world's preeminent experts on Egyptian pyramids, has stated that:

> "...*the pyramid was designed to be **a cosmic engine**...* (Author's emphasis.) *The mechanics of the pyramid as a cosmic engine depended on the Egyptian concept of a person and the distinct phases of life and death, called kheperu. These "transformations" continued when the ka, the ba, and the body, which had become separated at death, interacted in the final transformation—becoming an akh, a glorified being of light, effective in the afterlife. The pyramid was an instrument that enabled this alchemy to take place...*[31]
>
> *Joining the stars, the king becomes an akh. Akh is often translated as "spirit" or "spirit state." It derives from the term for "radiant light."*
>
> *The reunion of the ba and ka is effected by the burial ritual, creating the final transformation of the deceased as an akh. As a member of the starry sky...*[32]
>
> *The Pyramid is a simulacrum of both the mound of primeval earth and the weightless rays of sunlight,* **a union of Heaven and Earth** (author's bolding) *that glorifies and transforms the divine king and ensures the divine rule of the Egyptian household.*[33]

Plainly the goals of Nehebu-Kau and the purpose of pyramids as defined by Dr. Lehner appear to be one and the same: to unite the ka and ba of the soul, creating an akh, a glorious being of light. Having that in mind, this proposal and the significance of this 27.5-inch spinal-based measurement being incorporated in the physical construction of pyramid cosmic engines would not only be appropriate but also quite elegant.

More evidence in this proposal comes from a unit of measurement

Sacred structures connecting Heaven and Earth; the measurement results   51

noted in the hieroglyphs. The "aakhu meh" unit of measurement[34] was noted earlier in the possible link between the nbj/nb measurement unit to Nehebu-Kau, who unites the ka and the ba. When this union of the Egyptian soul occurs, it becomes an aakhu (akh or khu). As this is the believed purpose for the pyramids, it seems appropriate that the name of the unit of measurement of this cosmic engine would be related to such purpose. Had the ancient Egyptians in a nuanced transition from the measurement's physical purpose to its spiritual purpose also called it the aakhu meh, a spirit/light measurement? Dr. Robert Schoch in his book *Pyramid Quest* cites H.S. Lewis in the different determinations of the etymology of the word "pyramid," which was derived from the Phoenician "purimmiddoh," meaning "light measures."[35] The Phoenician civilization's time period closely paralleled the Egyptian civilization, and they were known for their extensive seafaring trading, including with Egypt. If the word "pyramid" did come from the cited Phoenician word "purimmiddoh," it provides a straightforward link for the use of the Egyptian "aakhu meh" measure with its meaning of a spirit/light measure. I find that meaning with the discussions here makes great sense.

This is definitely in the realm of possibility. I would include in the subtleties of the Egyptians in particular for the aakhu meh, their word used with their measurement names, the word "meh." I noted earlier Egyptologists have substituted the Latin term cubit (elbow) as a catch-all term in relation to their measurements, whereas the Egyptians used meh. I think the distinction of the use of meh is significant. Sir Flinders Petrie describes the meaning of the Egyptian meh:

The name of the cubit, meh, is also that of a binding, or girdle, or diadem; as it is just the length of a head-band for a smallish head (size 6 5/8), and it cannot agree with any defined part of the arm, it seems that the primitive measure was named from the head fillet.[36]

52   The Spirit of Light Cubit

Sir Petrie's meaning of meh fits specifically for the spiritual nuances of a spirit/light measurement derived from the spine and symbolizing the raising of spiritual energy of an enlightened consciousness. The meh seems to represent a binding (uniting) crowning headband with the symbolism signifying the raising of such subtle energy to higher consciousness and uniting Heaven and Earth within oneself. In using such terms as cubit to replace a culture's original terms, so much may get lost. Egyptian hieroglyphs work on many levels.

This archeological and physical evidence infer the solid possibility of the use of a 70-cm (27.5-inch) unit of measurement in the Old Kingdom pyramids. Now after this long but necessary evidentiary introduction of the existence of an Egyptian unit of measurement of 27.5 inches (70 cm), let us conclude this section with the results of using this measurement in the dimensions of the Great Pyramid.

Fig. 11 The Great Pyramid and Sphinx

Sacred structures connecting Heaven and Earth; the measurement results 53

Forgive me, just one more paragraph first. The Great Pyramid, the last remaining wonder of the ancient world, never ceases to awe and humble me. Whether seeing a picture of it or remembering, how fortunate I have been to visit it, stand next to it, even journey up into the queen's chamber, through the Grand Gallery, into the king's chamber, and even down to the subterranean chamber and the pit. Sometimes I can see it in my mind's eye when it was first completed, its capstone in place, covered with polished, and gleaming white Tura limestone shining in the Saharan desert sun like a star that has come to Earth. It was the height of a 48-story building with each side longer than two and a half football fields, and an architectural footprint of 13 acres that even today is less than an inch of being perfectly level. It was surrounded by a wall almost 30 feet tall. There's more but let that suffice for such a creation by human beings that seems almost impossible in its physicality but also should be experienced, felt, if you will. If you ever have an opportunity to visit the Great Pyramid, do not hesitate.

As already presented in Chapter 3, no matter what unit of length is used for the dimensions of the Great Pyramid, the proportional results for its perimeter over its height will result in 44/7; this is the mathematical constant 2 pi, the constant used in the formula to determine the circumference of a circle and the time period of a simple pendulum. Such results are curious and significant in themselves, as previously pointed out. Further, the height of the Great Pyramid over its width, reduced to its lowest terms, results in seven over eleven (7/11), and this result becomes significant when the 27.5-inch (70 cm) unit of measure is used.

In using the measurement of 27.5 inches the results are intriguing and thought provoking. The Great Pyramid's original height has been determined as 481 feet. When the Egyptian unit of measure, 27.5 inches, is used rather than a foot (12 inches), the results provide an answer of 210 units of nb (nbi, nebiu, aakhu meh, spirit of light, spinal cubits). Particularly, this result (210) is divisible by seven (7 x 30 = 210), which was also noted by Dr. Stecchini in his treatise as a necessary criteria put forth by the great polymath Sir Isaac Newton in his own study of the dimensions of the Great Pyramid. For transparency purposes, mainstream Egyptology maintains the "royal cubit" of 20.61 inches was used in the Great Pyramid, providing a height result of 280, which is also divisible by seven (7 x 40). Yet using a unit of 27.5 inches is compelling in its specific result of 210. Egyptologists believe that the Great Pyramid height consisted of 210 tiers (levels). The fact that a 27.5-inch unit and only such a measure results in a height result of 210 units outweighs coincidence and points toward deliberate purpose.

When the width of the Great Pyramid of 756 feet is translated into units of length of 27.5 inches, the result is 330 units, another interesting result. I correlate the side base length of 330 units to the human spine with its 33 vertebrae. This result augments the appropriateness for the use of this measurement and the presentation that it is representative of the human spine. This assertion is reinforced from archeological finds shared to this juncture, but also, specifically, as stated in Chapter 2, that the Giza Plateau was dedicated to the Egyptian deity Osiris, whose symbol was the djed hieroglyph, meaning "Osiris's spine." The ancient Egyptians even had a specific ritual of "Raising the Djed," a ceremonial rite which served not only as a metaphor for the stability of the monarch, but also symbolized the resurrection (rebirth) of Osiris. I see this rite incorporating the rise and rebirth of consciousness toward becoming a radiant being of light, an aakhu in life rather than death.

As a final observation on the physical, archeological, and numerological aspects, a fascinating mathematic feature that rests within the result that only occurs using a unit of 27.5 inches, which is a height of 210 and width of 330 such units. This mathematical feature involves prime numbers, those important and unique integer numbers that are only divisible by one and themselves. The first four prime numbers are 2, 3, 5, and 7. If you multiply these four numbers together (2x3x5x7), the product is 210! Remember I showed that the Great Pyramid's height over its width in its lowest terms is 7/11. So, continuing to the width of 330 units, the very next prime number after 7 is 11 and thus if you then multiply the prime numbers 2, 3, 5, and substituting 11, since the height used 7, the result (2x3x5x11) is 330! In prime numbers both the Great Pyramid's height and width share 2, 3, and 5, and its height over width proportion has already been shown to be 7/11, which brings one back to correlating and multiplying these primes to the dimensions 2x3x5x**7** = 210; 2x3x5x**11** = 330. Importantly this only occurs if a unit of measurement of 27.5 inches is used. These are unique results to this unit of measurement.

There is evidence of the purposefulness of these mathematical aspects presented above. Here is the incredible, elegant aspect of all this: the missing 25.4 feet in the Great Pyramid's height (the top seven courses) believed to represent the missing capstone, when converted into 27.5-inch units, results in 11! So the top seven horizontal courses equate to a missing vertical height of 11 27.5-inch measurement units, creating a proportion of seven horizontal courses to 11 vertical units of measurement. I find this astonishing in that the proportion of the original entire height of the Great Pyramid to its full width is 7 over 11! The dimensions and courses of both the Great Pyramid and their proportions of its capstone of 7 and 11 reflect and thus magnify each other. I believe the Great Pyramid and its capstone are purposely and perfectly designed to reflect and unite and magnify itself in such a significant and elegant manner. Just like the Hermetic axiom of "As Above, So Below," which represents not only the reflecting but the uniting in One. The proposal can be debated, but the mathematical results are there. It could be considered the capstone of the evidence being built here that has brought us to this apex.

In my opinion this provides strong evidence that such mathematics were purposefully designed in the Great Pyramid, particularly the prime numbers. As an example of such use, I will cite a present-day example—the radio message sent by the U.S. Arecibo radio telescope. Until recently this was the world's largest radio telescope used to plumb the mysteries of deep space, including seeking and studying pulsars, quasars, and other deep space radio phenomena. In 1974, primarily as a public awareness and relations event, these deep space researchers announced they were going to send a message from Earth into deep space in the possibility it could be received by other intelligent life in the universe. This began a debate on how to send such a message and make it unique and recognizable as being purposely sent by intelligent life. The conclusion was to send the radio message out in bursts of prime and semi-prime numbers, for in doing so, the radio transmission would be recognized as not a natural event or happenstance but intelligently and purposely sent.

Here there is evidence that the ancient Egyptians, using the proposed unit of measure, were leaving a mathematical message that, at least, their design was purposeful.

B: The Maya

In my journey after years of research into this neglected but evidentiary incredible Egyptian unit of measurement and finding such abundant and compelling evidence validating its importance, I couldn't help but wonder could this measure and what it symbolized have been used by other cultures at their sacred sites for similar reasons. As I mulled over these thoughts, I was first drawn to the ancient Maya civilization. I would venture to write that the Egyptians have the most well-known and famous pyramid building civilization and the Maya have the second. Hence, though an ocean away, I turned my eyes and thoughts in relationship to this measurement to the Maya pyramids and temples in Central America.

Mentally I was preparing myself for an extensive dive into Maya archeology similar to the depth and breadth of the dive into Egyptology. I do not mean this in any negative sense. There is so much value and learning in the journey itself, to learn more of these incredible cultures and travel to these sites and meet their descendants and others who kindly share their history. Also, along the way, I had the opportunity to engage with so many other travelers seeking in their own journeys. I learned and hopefully gained some wisdom throughout it all. Through all these incremental involvements, it felt as if I was experiencing the "Butterfly Effect" personally. In a very over-simplified explanation, the butterfly effect is a term which comes from chaos theory. Chaos theory proposes that there is much less randomness in the universe than that appears and that there are deep patterns to much that occurs. These deep and hidden patterns often come from small, sensitive initial events. There are limitations to this theory. In any case it felt like the initial events of my journey and research were leading to recognition of such patterns and a greater picture. The threads I was pulling together of this shared cultural tapestry were revealing more of the pattern and its picture.

The groundwork and foundation of this 27.5-inch unit of measurement had been done in the ancient Egypt research; I was beginning to see a pattern. This was a matter of transition from the sands of the Sahara to the jungles of the Yucatan. I was excited, curious, and at the same time expecting a prolonged journey and search as I had encountered in Egyptology. I was pleasantly surprised. The Maya researchers and archeologists had done this important arduous work already. They have researched and determined that the Mayans had a similar measurement for their pyramids called a "zapal."

Civil engineers Craig Smith and Kelly Parmenter performed an overview of the Mayan measurement system in an article entitled "An Ancient Maya Measurement System" from the January 1986 issue of *America Antiquity*. The article focused primarily on the principal dimensions of ten buildings at three ancient Mayan sites, including Chichen Itza.

Looking at common dimensions of the Mayan structures, the authors calculated a standard unit of measurement of 1.47 m (147 cm), with a variance of ±5 cm, called a zapal. As a test for the hypothesis, several principal dimensions of the Kukulkan Pyramid were converted into the zapal equivalents using the above unit of measure. Two of the primary measurements were the width of the top of the Kukulkan Pyramid and the temple erected on the platform on the pyramid top. It should be noted that the Maya culture built truncated pyramids that did not come to a point as the Egyptian pyramids did, but ones with a flat top to place a small temple upon. Two of the results of their study were a pyramid top of thirteen zapals across with a temple of nine zapals across.

The findings resulted in the numbers 9 and 13, significant numbers for the Maya. The number 13 was associated with the levels of paradise, and the number 9 referred to the levels of the underworld. As can be seen, these researchers also used results in meaningful numbers to validate a proposed unit of measurement as I did for Egypt.

Taking the variance of ±5 cm (142 cm) of a Mayan zapal and the comparison to the 70 cm (27.5 inches) Egyptian measurement, it is clear that two of the Egyptian units (140 cm) are approximate to a Mayan zapal. To use Dr. Stecchini's terms, the Mayans used the measurement of a full yoke rather than the Egyptian half yoke. I do note that 142 cm is exactly the length of two average spinal columns as documented earlier.

The Kukulkan Pyramid was dedicated to their god of the same name, meaning "The Plumed Serpent." This follows a similar theme from Egypt of winged or upright serpent symbolism. Notice it fits the concept of raising the kundalini through the spine, raising one's consciousness, symbolized as being winged or upright.

Fig. 12 Kukulkan Pyramid (El Castillo) at Chichen Itza

There also is an interesting and curious similarity with Mayan hieroglyphs and statues that depict a double-headed serpent bar, portraying a serpent with a head at each end of its body, perhaps illustrating the Maya zapal measurement of 55 inches (140 cm)—a double Egyptian 27.5-inch measurement. This double-headed serpent also sounds similar to the Egyptian serpent god Nehebu-Kau, who, as noted earlier in this chapter, would also be depicted as having two heads, one at each end of its body. One wonders at the similarities.

Along with the Egyptian pyramids and their deity Nehebu-Kau, the Maya Kukulkan Pyramid site is named for the Maya deity Kukulkan, the upright plumed/winged god, and the Maya double serpent bar represents axis mundis, points where Heaven and Earth come together. The Kukulkan Pyramid was aligned to connect with heavenly events, and the Maya serpent bar is symbolic of uniting Heaven and Earth.

"Terence Grieder was the first to propose that the ceremonial bar identified its holder as a personification of the "world tree" or the axis mundi, thereby directly connecting the holder to cosmic deities and the celestial realm."… The well-known homophony between the Yucatec words for snake (chan) and sky (caan) is an important part of this argument.

"…Freidel, Schele, and Parker considered the ceremonial bar to be both the ecliptic and a Heavenly umbilicus connecting the mundane with the celestial."[37]

With the importance and symbolic significance of the Maya double serpent bar, it follows well that rather than a measurement of one spine/serpent of 27.5 inches, the Maya with a double serpent iconography will have a measurement of 55 inches.

The archeological evidence showing the similarities of the Egyptian and Maya cultures from pyramids—through units of measurements continuing with serpent symbolism and creating axis mundis—is persuasive. If it were not for these cultures' separation by 7,600 miles, an ocean, and their times of antiquity, I think archeologists would believe there had been communication between them. I think it is time to revisit the possibility, as evidence is building.

In any case the Maya researchers have presented a unit of measurement used by the Maya engineers and architects of an approximate double spinal light cubit, which in the study of ancient metrology of cultures' comparable units of measurement, let alone purpose and symbolism, presents strong evidence of links.

C: The "New World" Ancestral Puebloans

This continuing journey remains in the New World following the trail from the Maya in Mesoamerica onward north through Mexico into North America as far as what is now New Mexico and Colorado. The primary focus will be on the pre-Columbian culture of the Ancestral Puebloans and related cultures. The Ancestral

Puebloans are a culture/society described by some archeologists and anthropologists as a member of "Oasisamerica," a term used to incorporate a broad area of several connected Southwestern cultures, including the Ancestral Puebloan, Mogollon, Hohokam, and Patayan, a smaller group toward California. See the following map.

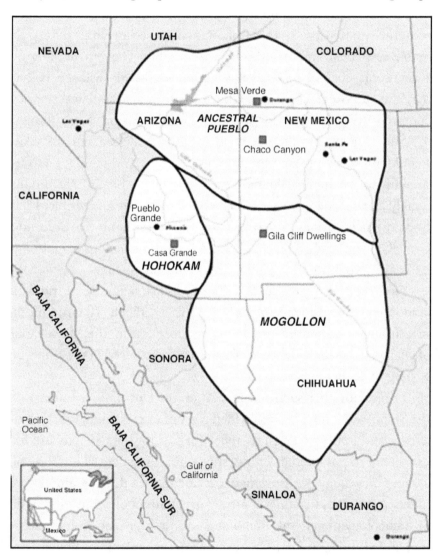

Fig. 13 Ancestral Puebloan area of culture

Personally trekking through such incredible sites as Chaco Canyon, Mesa Verde, Aztec Ruins, Salmon Ruins, and related sites in curiosity and admiration of these cultures, I found their great houses and kivas are a reward in themselves. I had the opportunity to camp for the weekend in Chaco Canyon National Historic Park in view of Fajada Butte, home of the Sun Dagger site (an archeoastronomy site that tracks both the annual sun events and multi-year moon events. An important aspect of the Sun Dagger marking of solar events, such as solstices and equinoxes, as you will see, is that they are **marked at noon.**) Chaco Canyon is an isolated site not compromised by any light pollution for star gazing. Even though the Sun Dagger and other Chaco structures' alignments are solar, I could appreciate gazing up at an evening sky and see the heavens similarly as the ancient Puebloans did, incredibly beautifully and infinitely star studded, with the Milky Way so bright and visible, I felt I could reach out and touch it. I would like to thank the Colorado Crow Canyon Archeological Center for their archeological camping tours and explorations to sites such as Chaco Canyon.

As so often in my research, as I was on such magnificent tracts of land looking for evidence of links for this 27.5 inch (70 cm) unit of measurement or multiples of it, like its close cousin, the Maya zapal of 55 inches, such evidence was also to be discovered in the tracts of writing of archeologists.

I found another such recent written tract by an archeologist of the Southwest, and admittedly I felt a pronounced validation, a resonance, and honestly a sigh of relief for this journey. If may ask for the reader to put oneself "in my shoes" for a moment. The study of ancient metrology (measurements) is not an area that is glamourous, highly promoted, nor does it attract great attention from the public or archeology in general. I felt like I was in "terra incognita" a solitary and trepidatious place, yet exciting at the same time. Then as my research continued I discovered I was not alone, though there were not many. I found writings of scholars and members of academia pursuing the important research of ancient metrology. As shown in the previous chapters from such scholars and the evidence presented here, it is an important area to study to aid in understanding the levels of consciousness of cultures, their advancements, along with the connection and communications with other cultures. I am eternally grateful for these unsung academics and researchers for their work.

Walking in my shoes started with an idea that a unit of measurement of 27.5 inches (70 cm) that I had read about being used in ancient Egypt. Initially I couldn't even find such a measurement documented for Egypt. Fortunately, I continued the dogged research, finally finding archeological documentation for this measure's use in Egypt. From this point, along with pondering its origination and symbolism, I began to plug it in to multiple sacred sites of different cultures to see what the results would be. The consequences of this exercise provided results that were curious, interesting, and rousing enough for me to continue the pursuit of the possibility and significance of this unit of measurement being shared much further, wider, and earlier than anyone would expect, and exhibiting even more profound ramifications, a possibility that has not been put forth anywhere else to my knowledge. Then, further on, was the report from Maya researchers identifying that culture's unit of measurement, named a zapal, and estimated at approximately 55 inches (140 cm). Most recently I discovered an academic article on North America's pre-Columbian culture from the Southwest earlier identified as the Mogollon culture. It was a treatise of their units of measurement. The article's conclusion was that they used a unit of measurement of 70 cm—yes, a measurement of 27.5 inches! My shoes feeling worn by this journey now felt as if they had been re-soled, and my spirit felt "re-souled" as well.

Shared below are excerpts from research discoveries of another archeologist's written tract validating the use of such measurements put forth in this book's proposal entitled: "Archeological Metrology: A case study from Paquime." (Author's note: Paquime is also known as Casas Grandes and considered as part of the Mogollon culture. Located in northern Mexico near the borders of New Mexico and Arizona, it is one of the largest and most complex Mogollon sites in the culture's scope.) The article's support of a 27.5-inch (70 cm) measurement begins in its abstract:

"Using the method of Quantum Clustering, we find evidence for a measurement unit of 70 cm that is reflected in domestic and religious architecture."[38]

This paper on Paquime's discovery was another "oh my gosh" eureka moment. Remember what archeologists have written about cultures sharing the same unit of measurement or its multiples? Either they had strong communications with each other or the unit of measurement came from the same source. This provides more validation for my premise, and the ramifications are stunning. There it is, right in the report title—the same unit of measurement in Egypt and a derivative of it in the Maya culture. This report also included the measurement evaluation of multiple cultural ball courts in the surrounding region, specifically stating: "Expanding the evaluation to additional ball courts further indicates consistency with a unit around 70 cm."[39] (Author note: 27.5 inches.)

My research also found their case validating Paquime's use of a 70-cm unit of measure is further strengthened by a separate report in 2015 on the Albert Porter Pueblo site located in Southwestern Colorado; their research determined that the average diameter of kivas approximately 3.5 m in diameter. This converts to 5 units of a 70 cm unit of measure, providing another authenticating point for a measurement of 70 cm use by these cultures.

I also include from this report the excerpt that echoes the overall premise put forth by archeologists presented here, as they also use Sir Flinders Petrie's observations on the importance of metrology toward understanding the depth and breadth of cultures.

"As early as Sir William Petrie (1879), archeologists argued that metrology could provide valuable cross-cultural data and show historical relationships, but metrology studies tend to be infrequent and limited in scope. They haven't been placed within general anthropological framework, which is unfortunate *given that metrology impacts and reflects a culture's cosmological,* economic, and technological structure.[40] (Author's emphasis.)

Before finding the article on the 70 cm Paquime measurement, I had already been researching this measurement in relationship with the Chaco Canyon, Aztec Ruins, great kivas, and Salmon Ruins' diameters and finding collaborating results at these sites. There is also independent verifying evidence for connections between these sites from a different approach by archeologist Dr. Stephen Lekson. He has put forth a theory that connects the people of Chaco Canyon, New Mexico, to Aztec Ruins, New Mexico, approximately 60 miles north of Chaco Canyon and south approximately 500 miles to Casas Grandes/Paquime, Mexico. His theory—based on the fact that these sites are in a north-south longitudinal alignment he named the "Chaco Meridian"—is that these peoples were fixated on this alignment for ceremonial purposes in their architecture. Dr. Lekson further supports the social and ceremonial link of these sites using the evidence of macaw remains, T-shaped doorways, and other shared aspects. I believe my research assists his theory as his evidence supports mine. I think it is especially true with my research at Aztec Ruins National Monument in New Mexico.

In my own research of some of the remaining ancient kivas using measures such as 70 cm and validated by the Casas Grandes/Paquime use of 70 cm or this measure's Mayan kin of 1.4 m, the results are validating and statistically significant, averaging less than a 2 percent difference. They are listed below.

- Casa Rinconada great kiva; Chaco Canyon, New Mexico: Interior diameter: 19.3 meters: Equals 14 (13.8) zapals or 28 spinal/light units (27.5). I have called this measurement the spirit/light cubit but I do not want the significance of the spinal symbolism to be forgotten either.
- Pueblo Bonita great kiva I: Chaco Canyon, New Mexico: Interior diameter: 15.6 m: Equals 11 (11.14) zapals or 22 spinal/light cubits (22.3).
- Salmon Ruins great kiva: Interior diameter: 14.5 m: Equals 10 (10.35) zapals or 20 spinal/light cubits (20.7). I would like to thank Larry Baker, the executive director of Salmon Ruins Museum who was another one of my guides escorting me to Chaco Canyon.
- Aztec National Monument great kiva: Interior diameter: 41' 3.5" (12.58 m): Equals 9 zapals (8.99) or 18 spinal/light units (17.97).

(Author note: Kivas were and are an important aspect of Puebloan culture. They are considered to have a variety of teaching, social, ceremonial, and spiritual purposes.) This seems appropriate for a sacred structure designed to be an axis mundi, where traditional lore to live on Earth is united with religious tenets to live in spirit. To quote Dr. Susan Ryan, Crow Canyon's director of archaeology: "Architecture is the place where the ethereal and non-material qualities of the **cosmos were interpreted by ancient architects and emphasized in material form**. Architecture communicates culturally prescribed, and accepted, information to the observer about Pueblo cosmology."[41]

I believe the great kiva at Aztec National Monument in particular is a wonderful example of Dr. Ryan's observations, visitors are allowed to enter this magnificent, reconstructed great kiva, and the diameter measurements using units as shown above are incredibly accurate. The measurement results of 9 or 18 using the units of measurements above are also culturally significant and what I will call a fortunate, serendipitous event that occurred on my visit may provide an answer to one of the great kiva's uses.

Fig. 14. Aztec National great kiva excavation photo.

Fig. 15 Great kiva reconstruction today

Image 14 is a picture taken from the Aztec Ruins guidebook showing the early excavations of the great kiva, and image 15 is the completed great kiva reconstruction (personal photos).

The great kiva diameter results of 9 and/or 18 units (dependent upon using 27.5- or 55- inch units) reminded me of two important aspects: first as noted earlier at the Maya Kukulkan Pyramid, one of the measurements results of a zapal was 9—a significant number to the Maya and considered a sacred number for the ancestral Puebloans. The second aspect being the archeoastronomy discovery called the "Sun Dagger" site in Fajada Butte located in Chaco Canyon and a possible relation to Aztec ruins.

Thanks to the discovery and groundbreaking research of Anna Sofaer and associates at the Sun Dagger site and in Chaco Canyon architecture, we know that this site was designed not only to mark the solar equinoxes and solstices, but also to mark the much more involved cycle of lunar major and minor standstills: an 18.6-year cycle, 9.3 years from lunar major to minor standstill and 18.6 years to return. Besides their Sun Dagger discoveries, Sofaer and associates' research has found that 12 of 14 major buildings located in and around Chaco Canyon are oriented to solar and lunar positions and alignments. I would highly recommend reading *Chaco Astronomy* by Anna Sofaer; it is well worth it for the research and her own efforts. As an aside, on one occasion while I was visiting Chaco Canyon's visitor center and bookstore—so often the best research books are at the actual sites—I met Ms. Sofaer. I was so tongue-tied, I could only stumble over thanking her for all her work and never asked any questions; sadly, a missed opportunity.

The Aztec Ruins great kiva appears to have been designed for similar purposes as the Sun Dagger site. As pointed out earlier, the diameter of this kiva—9 Maya zapals or 18 Casa Grande/Paquime units, or as I have also call them spinal/light units—matches, in whole numbers, the lunar cycles tracked at the Sun Dagger site.

Sacred structures connecting Heaven and Earth; the measurement results   69

Fig. 16 Fajada Butte with Sun Dagger site circled

My serendipitous discovery when visiting the Aztec Ruins great kiva relates to the kiva marking annual solar events. I first visited Aztec Ruins National Monument on May 31, 2015. This is another superb site that is well tended by its caretakers. As I read the guidebook and explored the site, taking photos with my focus toward my research, I tried to just "be" there and absorb it *in toto*. I think readers can appreciate that sometimes when we travel to different destinations; it often seems we have to look at our photos to see where we have been. It was only when I returned home and was reviewing my photos that I realized that while I was there, staring me in the face, I was seeing a solar event in the great kiva that appeared to evidence it was designed to mark the summer solstice and possibly also the winter solstice. (See below.)

Fig. 17 Aztec great kiva; light beam on left vault

This photo of the great kiva was taken the 31st of May 2015 in a north-looking perspective. On the left-hand side (west) of the photo, you see one of the two kiva vaults, also called foot drums. Both the kivas and vaults generally have north-south alignments. Archeologists are not sure of the purpose of such vaults/foot drums, but believe they are related to ceremonial reasons. You should notice on the left side that a shaft of light is landing on the upper corner of the vault, illuminating it. This shaft of light comes from the roof hatch of the kiva and is its original entrance. This date is just three weeks from the solar summer solstice, which occurs on June 21, 2015. I believe part of the purposeful design of great kivas was to mark such events in some manner.

To better confirm this possibility, I was able to return to the great kiva in the fall to see where the shaft of light would fall. Please see the following photo.

Sacred structures connecting Heaven and Earth; the measurement results 71

Fig. 18 Aztec great kiva; light beam on right vault

This later photo of the great kiva was taken the 27th of October 2015, again in a north-looking perspective. Notice that the shaft of light is now on the opposite side of the kiva from the first photo above, now on the right-hand side (west) of the opposing vault. This is about eight weeks prior to the winter solstice on December 21, 2015. It appears that great kivas may also have functioned, similarly to the Sun Dagger, at least to mark the solstices. Unfortunately, I was unable to be present at Aztec Ruins on the actual solstices, but I did call the park rangers, asking if they could confirm on those days that the shaft of light illuminated the vault/foot drums. In each case they said it did. I greatly appreciate their kindness, for I am sure this was a strange call, with an even stranger request. I thank them sincerely and hope they were not just humoring an odd call and request.

On one of my visits to the great kiva at Aztec Ruins, one of the kind park guides was giving a history of its reconstruction and stated that after the reconstruction, elders from the descendants of the ancient Puebloans were invited to examine and give their observations on the reconstruction. The guide stated they thought it was a good reconstruction, but the roof was the incorrect height. The roof height would affect where the light, whether sun or moon, would land inside the great kiva. It makes me wonder about the remains of many great kivas that have been found with their roofs burned off. Was this done purposely to keep the solar and moon tracking knowledge concealed? The Puebloans documented knowledge from Chaco Canyon's Sun Dagger site and Chaco's building alignments of marking solar and lunar events. The possibility that Aztec Ruins' great kiva and others were designed to mark these in some fashion is quite real.

I follow these observations and photos with an artist's depiction of how such great kivas could function also as a solar seasonal calendar.

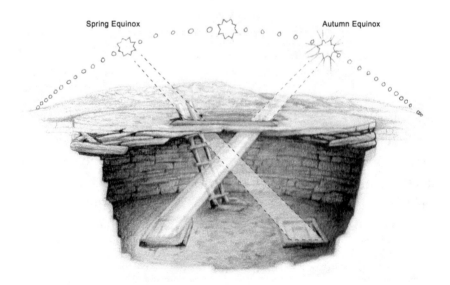

Fig. 19 Artist's depiction of a kiva marking solar seasonal events.

The concept of marking such heavenly events in such a manner is not without precedent or similarities. The Pantheon a sacred temple in Rome, Italy. It is over 2,000 years old and has some utterly amazing similarities to the Aztec Ruins great kiva. The Pantheon has a north orientation, which is considered unusual for European sacred structures. The great kiva has a north orientation. The Pantheon is primarily a round building with a domed roof and an oculus (opening) on its top (roof), comparable to the great kiva being a round building with a roof opening/entrance (oculus) and in both cases providing the only natural light source. The Pantheon has a rectangle portico on its north side; the great kiva has a rectangular room with an entrance on its north side. The Pantheon has 14 blind windows in its upper courses. The great kiva has 14 wall slots in its upper courses. Just in the physical descriptions of these two sacred sites, this is an impressive list of curious similarities.

The architecture of the Pantheon and the great kiva has been studied in depth, but neither's purpose or function is well understood. A recent academic paper on the Pantheon may, dare I say, shed more light for both.

Scholars Hannah and Magli have published a recent paper that does just that: "The role of the sun in the Pantheon's design and meaning."[42] Briefly, their conclusion is that one of the Pantheon's functions is similar to a specific type of sundial called a hemicyclium, described as a globe-shaped sundial with a hole in it that aimed a ray of light into the concave interior to markings that notated seasonal solar events, not to act as a daily sundial. Such notations "represent the diurnal passage of the sun, usually on four particular occasions (author's note: solar solstices and equinoxes)…but rather to substantiate *the symbolic connection of the building with the path of the sun* in the course of the year … the sunlight beam at *noon* is always located on a *meridian line*" (author's emphasis).[43]

One of the important aspects to be able to mark these four solar occasions was a northern alignment to create a north-south meridian (a line of longitude) so the ray of sunlight would function as described, specifically at noon. The Pantheon and the Aztec great kiva are both on such north/south alignments, and my experience at the Aztec great kiva points to noon as one marking point. Remember earlier in this section of Chapter 4, it was shared that archeologist Stephen Lekson theorized a north/south "Chaco Meridian" linking Aztec Ruins, Chaco Canyon, and Paquime/Casas Grandes; a similar postulated meridian line for the Pantheon helps to explain such a meridian "fixation" on these Puebloan sites. I also noted in Chapter 3 that the Chaco Canyon Sun Dagger site marked the solar equinoxes and solstices event at noon, and the Pantheon's solar calendar also functions at noon. There is also the coincidence for the determination of the length of today's meter discussed in Chapter 3. It was a meridian to be measured for the meter to be chosen as an international meridian versus a meridian primarily through one country. I wonder if the Puebloans had a similar discussion; perhaps the focus on a meridian alignment for ceremonial purposes was linked to more than tracking solar seasonal time events, and the measurement of the kiva's diameters can provide a symbolic clue.

Before leaving this section, we need to address the correlation of the historical, spiritual characteristics of kivas and the proposed relationships of the spinal/light unit of measurement. Using the knowledge provided by the Puebloans descendants, the Hopi proposed a symbolic connection between the skull and the spine upon which the head rests and which protects our central nervous system—the spinal cord and brain.

In Chapter 2, in relating the length proposed here being derived from the length of the spine, I quoted from Frank Waters, *The Book of the Hopi*. Below I have included an extension of that quote which includes its reference to the human skull and, in infancy, the fontanelle (soft spot).

"The living of man and the living body of earth were constructed in the same way. Through each ran an axis, man's axis being the backbone, the vertebral column... Along this axis were several vibratory centers which echoed the primordial sound of life throughout the universe...The first of these in man lay at the top of the head. Here, when he was born, was the soft spot, 'kopavi', the 'open door' through which he received his life and communicated with his Creator."[44]

The kopavi, the open door at the top of the head, was a vibratory center attuned to the creator and could give spiritual guidance to the Hopi. This is exampled in several Hopi traditions when groups were losing their way in physicality and materialism. For those who stayed on the spiritual path, the kopavi would guide them away from such "falls" to places of emergence to a new world. The Hopi tradition has human beings presently in the fourth world.

Mr. Waters' book continues, showing the Hopi philosophy combining the physical and spiritual connecting with God in the material world through the spine and raising the energy up through the crown of the head.

"Man is created perfect in the image of his Creator. Then after 'closing the door,' 'falling from grace' into the uninhibited expression of his own human will, he will begin his slow climb back upward. ...With this turn man rises upward, bringing into predominant function each of the higher centers. The door at the crown of the head then opens and he merges into the wholeness of all Creation, whence he sprang...that he may know himself at last as a finite part of infinity."[45]

The kopavi—the open door on the crown of the head which helps guide the Hopi to the site of raising themselves and their consciousness into the next world—also provides apparent symbolism relating to the spine. This proposal is based upon the Hopi story of their journey of emergence into the upper worlds. There are several versions of this story and generally they include climbing up or through a giant hollow reed (like bamboo) high into the sky to emerge into the next world through the sipapuni or sipapu, its entrance. Generally great kivas have such symbolic holes in their floors. It is believed some Hopi clans designate a geologic dome located in the Grand Canyon by the Little Colorado River upstream before it joins with the Colorado River as their place of emergence into the present (fourth) world. Though many kivas have flat roofs, the geodesic dome provides an archetypal representation symbolized by kivas. When I had the opportunity to visit the present Hopi homeland on the three mesas, I saw such a dome-like kiva resembling the Grand Canyon geologic dome.

From this evidence and research I propose that it appears that kivas, particularly great kivas, are, among their multiple purposes, symbolic representations of raising spiritual consciousness. It is not a far reach in visualization or symbolism to view the hollow reed as the spine raising them up to a higher world, then entering it through the sipapuni or skull foramen magnum into the higher world or consciousness seated in the skull. To continue the upward journey, the kiva roof opening as the kopavi, top of the skull fontanelle, the connection to the Creator consciousness. This is very sophisticated and straightforward for both the physical and spiritual. Provided here are images that should assist in the visualization of this concept.

Fig. 20 Geodesic dome/Kiva

Sacred structures connecting Heaven and Earth; the measurement results   77

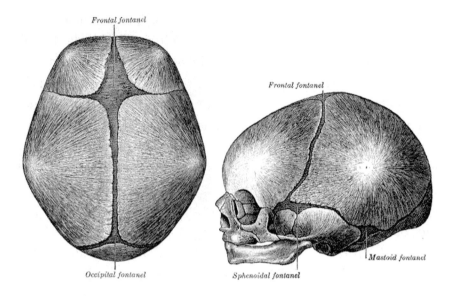

Fig. 21 Top of skull with fontanelle and side view with fontanelle noted on top.

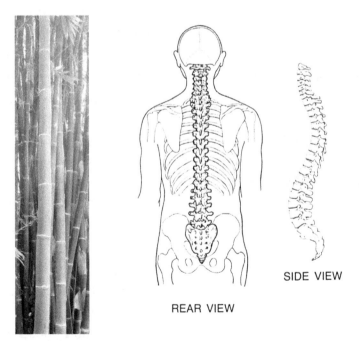

Fig. 22 Bamboo reed   **Fig. 23** Spinal column

I think the Hopis' spiritual philosophy and these visual aids provide solid credence to the proposed symbolism. The similarities of other cultures' practice of raising their spiritual energy or conscious (kundalini) through the path of the spine and up the central nervous system is evident in the creation and emergence stories from the Puebloan descendants, the Hopi. The Hopi creation and spiritual evolvement philosophy completes my proposal of this symbolism of great kivas.

Prior to going to the next and final section of this chapter, I want to assure Hopi society and their Elders there is no intent by this author of disrespect with the premises presented here. The Hopis' cultural and spiritual philosophies both have great beauty and depth. They are, understandably so, a private people. This research and theories came from the materials available and an attempt to understand an encompassing unity.

The evidence from the Archeologists at Paquime, Casas Grandes, providing their conclusion of the use of a 27.5-inch (70 cm) unit of measurement and Paquime's connections both with other Mesoamerican cultures, such as the Maya, along with Ancestral Puebloan cultures to their north provide further confirming indication of this unit of measurement and its multiples being shared in measure and symbolism by manifold cultures.

D. Stonehenge; Great Britain

This journey now travels back across "the pond," the Atlantic Ocean, alighting at Stonehenge on the Salisbury Plain of England. Stonehenge is an enigmatic megalithic site whose origins trace back to around 8,000 BCE, with evidence of tree postholes. Its upright stone circles are believed to have been erected at approximately 2,500 BCE, a similar time Egyptologists date the Great Pyramid, while the Z and Y postholes are dated around 1600 B.C. Overall the work of Stonehenge "proper" is thought to have been done over

Sacred structures connecting Heaven and Earth; the measurement results   79

Fig. 24 Stonehenge

a period of 1,500 years. The site is thought to have been used for not only ceremonial and spiritual purposes but also for astronomical ones. Though many of its possible astronomical alignments are debated, Stonehenge's summer and winter solstice alignments at sunrise and sunset are agreed upon. This makes it another axis mundi, a sacred place that connects Heaven and Earth.

Approaching Stonehenge is an experience in itself. It appears dramatically out of seemingly nowhere in the English countryside, seemingly mute yet dramatic, keeping its secrets to itself. There are the remains of two megalithic stone horseshoe shapes surrounded by similar remains of two stone circles. Archaeologists have also found remains of more concentric circles emanating out from the center up to the circular berm that encloses the site. The site is incredibly popular with many visitors, and it is amazingly quiet even so, as if everyone senses entering into a sacred space. Having had the opportunity to be "inside" the rings of Stonehenge, I can share with you the feeling is even more pronounced. Any sound seems to follow you.

Figs. 25 and **26** Stonehenge Solar alignments

Who built it is another mystery, particularly because of its many phases of construction over millennia and scant cultural evidence discovered. Some theories promote local Neolithic cultures with its construction, while others include ancient societies from Wales from where the "blue stones" (Preseli stones) were transported from. Recently a new theory by archeologists has been academically presented that traces Stonehenge's design and construction to an ancient European seafaring culture. This is a very exciting finding, particularly in assisting the premise of this book of an ancient unit of measurement used by multiple cultures over vast distances. It will be expounded upon in more detail at the end of this section. For now let us focus on this proposed shared unit of measurement use at Stonehenge.

There is great debate over the megalithic structures and stone circles in Great Britain and if they were created with a codified unit of measurement or not. This dispute of whether these incredible megalithic sites spread across the British Isles possessed or not a measurement system seems very odd to me. That these ancient unknown builders had the technical abilities to build such sites with advanced engineering, architectural and astronomical knowledge; yet they may not have had a codified measuring system appears farfetched. The knowledge of the measurement system may be a mystery now, but I believe it existed. In the search for this ancient measurement system it appears the one closest to finding such an answer—an answer considered still today very controversial—is Alexander Thom.

Alexander Thom was a Scottish engineer and the chair of engineering science at Brasenose College, University of Oxford. He spent over 40 years, with his expertise, scientifically measuring, with the eye of an engineer, over 600 ancient stone circles, primarily in Great Britain but also in Europe. From his measurements, research, and analysis, he determined that these ancient stone circle builders had developed a system of measurement for their construction and design. Basically Alexander Thom focused on three units of measurement from his body of work: the megalithic inch (MI) equivalent to 0.817 inch; a megalithic yard (MY) consisting of 40 MI then being 32.68 inches; and a megalithic rod (MR) consisting of 100 MI being 81.7 inches.

Thom's megalithic rod of 81.7 inches caught my attention. I had already been on my journey for some time in researching the light/spirit cubit of 27.5 inches and had been digging into volumes of written material on the dimensions of Stonehenge. I was beginning to find some preliminary fascinating results. When I read of Thom's work of a lifetime and his determination analyzing these megalithic measures, I could tell his megalithic rod (MR) was very close to three light/spirit cubits of 27.5 inches (82.5 inches). Resorting to pencil and paper and a calculator, I compared the megalithic rod to three light/spirit cubits, and the results were almost identical. The difference between them was only one percent—a single megalithic inch! The results were incredible and suggested a significant correlation. I also had to muse to myself that using Thom's megalithic inch, essentially 33 MIs equaled a spinal/light cubit proposed here and that 33 was also the number of vertebrae in the spine. Coincidence?

I looked into Thom's work further, specifically for his proposal for the use of his megalithic rod (MR), which he had determined from these Neolithic stone circles. The depth and breadth of Alexander Thom's work is extraordinarily detailed, and voluminous and justice cannot be done to it here. So I will only note Thom's suggested purpose of the megalithic rod. Thom thought that the megalithic rod was used for measuring larger circles and distances in particular for circles' perimeters. While he measured perimeters in megalithic rods, he measured circle diameters in megalithic yards (MY). He also believed these megalithic architects preferred design results in whole numbers. Thom determined the megalithic rod to be two and a half megalithic yards (100 megalithic inches).

I had noted earlier Thom's megalithic rod is an ideal match for three light/spinal cubits— similar to an English yard (36 inches) is comprised of three feet of 12 inches—and I knew this was my own cultural bias. Nonetheless I went forward to measure the diameter of Stonehenge's circles using a 1/3 portion of the megalithic rod described herein as a light/spirit cubit of 27.5 inches. The results were so surprising I did them over and over again to make sure.

Finding exact measurements of the Stonehenge circles is difficult and varies according to the use of the inside, outside, or mean diameters. The age and condition of remains at the site also hinder exact measurement of the original diameters. This has been a challenge for all the investigators and surveyors of such sites, and even so the results were startling. The first inner blue circle has been estimated at diameters of approximately 75 to 76.1 feet. Using the mean of these measurements, 75.51 feet, it represents almost exactly 33 light/spirit cubits (32.95). This was a very surprising and promising start considering the proposal that the unit of measurement of 27.5 inches derived from the length of the human spine, and the spine, as just noted, consists of 33 vertebrae. The next Stonehenge circle (the Sarsen stones) has a mean diameter of 100.80 feet, equaling 44 light/spirit cubits. The results surprised me again and I began to wonder if this was a pattern. Continuing on the multiple circle measurements at Stonehenge, the Z-postholes with a diameter of 127.87 feet translate to 55 light/spirit cubits (55.8), and the Y-postholes with a diameter of 177.16 feet translate to 77 light/spirit cubits (77.3). These outcomes are absolutely incredible and will occur only when using a unit measure of 27.5 inches. The results yield concentric expanding circles, measured by the light/spirit cubit, of 33, 44, 55, and 77. These results are straightforward and dramatic and go beyond what anyone could call coincidence. This showed a clear pattern, yet admittedly I felt some consternation: Where is the 66-unit diameter? Fortunately as I pondered and plugged along in the research, I believe I have found the answer to this nagging question. If the following diagram of the Stonehenge site is examined, it can be seen that there is marked a circle of unexcavated postholes halfway between the 55 light/spirit cubit diameter of the Z-postholes and the 77 light/spirit cubit diameter of the Y-postholes. I believe that this unexcavated circle of postholes between the Z and Y holes will, if examined, equal 66 light/spirit cubits. The results are clearly and obviously staggering and beyond coincidence. The diagram on the following page illustrates concentric circles at Stonehenge with light/spirit cubits of 33, 44, 55, 66, and 77.

Whereas the sites and cultures examined in the first three sections of this chapter all have cultural and academic documentation of a unit of measurement of 27.5 inches (70 cm) or its multiple in the case of the Maya, Stonehenge does not. I write this knowing I have just presented all the work of Professor Thom. It is because the general archeological community, sadly, has refused to entertain, let alone accept the conclusion of Professor Thom and have relegated it to fringe or pseudo-science. I find this very unfortunate particularly with Professor Thom's academic pedigree, scientific method, and lifelong dedication to this research. Hopefully, the research presented here in some small way will help to have his work reevaluated. In the case of Thom's work in relation to the proposed unit of measurement of 27.5 inches (70 cm), it exhibits a relationship of a multiple of three rather than the Maya measurement example as a multiple of two and using the light/spinal measurement provides significant results. Lastly, as I have noted earlier, some have argued a length such as 27.5 inches (70 cm) could be attributed to the length of a pace or step. A pace is not a body proportion, going back to what was written by the Roman engineer Vitruvius that body proportions are represented by the body, such as the width of a hand or the length of a foot or the distance from fingertip to elbow or, in my opinion, the length of the spine. I could see a difference in the use of

Sacred structures connecting Heaven and Earth; the measurement results   85

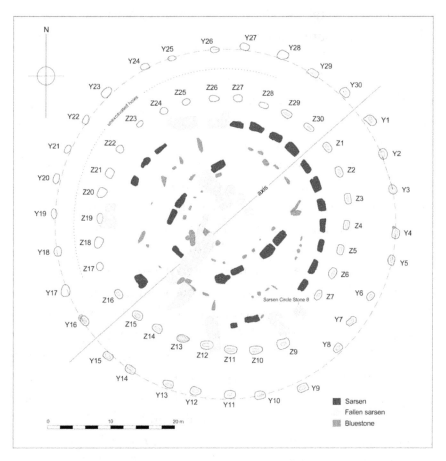

Fig. 27 Stonehenge circles diagram

Blue Circle = 33 light/spirit cubits
Sarsen Circle = 44 light/spirit cubits
Z Post Circle = 55 light/spirit cubits
Unexcavated Circle = 66 light/spirit cubits?
Y Post Circle = 77 light/spirit cubits

measures in civil structures versus sacred structures with the possible use of a pace or step in civil structures. While in sacred structures the adherence to the use of body proportion as Vitruvius identified, what better measurements (body proportions) for a sacred site than those created by the Divine.

Stonehenge, like the sites reviewed in the other three sections of this chapter, is also an axis mundi center, uniting Heaven and Earth, a site of spiritual and ceremonial use. I have also postulated that such sites were to raise human consciousness to connect with the Divine. Recent research and evidence at Stonehenge seem to clearly bear this out. The first research material I discovered was done in 2009, "The Sounds of Stonehenge." Dr. Rupert Till was the lead researcher and is a senior lecturer in Music Technology at the University of Huddersfield. Dr. Till was investigating the possibility that Stonehenge had a design incorporated into it for purposeful use of sound and music with a focused acoustical resonance at the site. He and his team's research led them to creating computer digital models to test the acoustics of Stonehenge that resulted in confirmations of their hypothesis. This led them from Great Britain to the city of Maryhill, Washington, in the United States, where a full-size, complete replica of Stonehenge—its stone construction and circles, using concrete—was located. The results and conclusions are fascinating and are a significant contribution to Stonehenge possibly functioning as a site to raise consciousness as offered here. I include below one small excerpt of Till's conclusions that point to the use of Stonehenge acoustics in creating meditative states:

"Those in the **very centre may have found their brainwaves becoming slowed to an alpha rate** (Author's emphasis) of around 10Hz, relaxed but alert. **Around the outside of the circle the slower rhythm would entrain their brains to be synchronized at around 5Hz to a theta wave state, becoming deeply entranced.** (Author's emphasis.) Some may have had epileptic fits triggered. The extended periods of time in alpha or beta wave brain states may have had physical effects, aided concentration, changed mood, induced visions, and may have been as healing activities."[46]

Till's team's conclusions noted that at the center of Stonehenge, the acoustics could have put an individual into an enhanced alpha wave state of meditation, and other individuals lining the stone circle itself could have been put into a much deeper theta wave meditation. As they point out, the leader or sacred elder could have led an initiation or healing type of ritual in a state of lighter meditation, guiding the "initiates" who are experiencing a deeper meditation into a higher or deeper spiritual consciousness. Please pause to think about this, as I did. What a wonderful validation of the idea that all these sites discussed had included in their purposes, construction, design, and use of measurement the raising of our consciousness through a meditative state. It reinforces my faith and hope in humankind.

For proper context and respect for Dr. Till, I include caveats for their acoustic findings at Stonehenge. I would also recommend going to their website to review their corpus of work on Stonehenge. You will not be disappointed.

"This project in no way claims to have decoded Stonehenge or explained its meaning and purpose; it does not suggest that Stonehenge was created as an outdoor concert or dance hall, music venue or amplifier. It does suggest that music may, alongside other visual, astrological, ritual or cultural elements, have had a part to play, and that this is worthy of investigation. It hopes to show that music, sounds and acoustics were likely to have been an important part of this iconic site."[47]

Almost concurrently with Dr. Till and his team, another group of researchers was doing similar research, presenting their results online in 2013 and in print in 2014: "Stone Age Eyes and Ears: A Visual and Acoustic Pilot Study of Carn Menyn and Environs,

Preseli, Wales, by Paul Devereux & Jon Wozencroft."[48] This pilot study, started in 2006, was supported by the Royal College of Art, London, to investigate the possibility of stones at different Neolithic sites that were chosen and dressed with acoustical qualities in mind, with a musical function producing sounds, when struck, similar to drums, bells, and gongs. In 2013, their team was able to test the theory at Stonehenge with the actual stones—tests that confirmed the stones' acoustical qualities.

Here are two different studies concluding that the sound-producing and acoustical aspects at Stonehenge are seemingly done purposefully in design. Imagine being there on the Salisbury Plain over 4,500 years ago at Stonehenge for such an experience, not only hearing the call of the Stonehenge "sound" rolling over the plain but participating in the meditative experience of a deep theta state; it's just breathtaking.

But wait; there's more! I know that sounds like an info-commercial, but it is true. There is more. There has been very recent and very exciting research, particularly for the premise in this book. Earlier in this section on Stonehenge, I wrote that very little is known of Stonehenge's builders or culture, and the standard theory has revolved around local inhabitants and peoples from Wales. This new study shows evidence for and puts forth the theory that it was built by a seafaring culture from Europe.

Dr. B. Schulz Paulsson very recently published her research: "Radiocarbon dates and Bayesian modeling support maritime diffusion model for megaliths in Europe."[49] Dr. Paulsson's multiyear study presents a massive amount of research that in its conclusion suggests that European megalithic structures, including Stonehenge, did not develop independently but rather originated from the Brittany area of northwest France starting around 5,000 BC and spreading out from there and including Great Britain. As excerpted in Dr. Paulsson's paper:

"Northwest France is, so far, the only megalithic region in Europe which exhibits a pre-megalithic monumental sequence and transitional structures to the megaliths, suggesting northern France as the region of origin for the megalithic phenomenon.

"A fresh expansion occurred during the first half of the fourth millennium cal BC when thousands of passage graves were built along the Atlantic coast of the Iberian Peninsula, Ireland, England, Scotland, and France. Their distribution emphasizes the maritime linkage of these societies and a diffusion of the passage grave tradition along the seaway." (Author's emphasis.)

"The megalithic movements must have been powerful to spread with such rapidity at the different phases, and **the maritime skills, knowledge, and technology of these societies must have been much more developed than hitherto presumed**. (Author's emphasis.) This prompts a radical reassessment of the megalithic horizons and invites the opening of a new scientific debate regarding the maritime mobility and organization of Neolithic societies, the nature of these interactions through time, and the rise of seafaring."

Dr. Michael Parker Pearson, an archeologist and Stonehenge expert, supports the conclusions, stating, "This demonstrates absolutely that Brittany is the origin of the European megalithic phenomenon."[50]

What her results suggest with this proposal of a seafaring culture spreading and sharing the skill, technology, and expertise to construct such Neolithic sites and going back almost 7,000 years is multifold. Not only was this ability passed on and shared with other cultures, but it strengthens Professor Thom's belief that a shared unit of measurement was also used and passed on along with the cultural/spiritual purpose I have proposed. The evidence also demonstrates that Neolithic seafaring was thousands of years more advanced than thought. Such ancient seafaring, more ancient than expected, opens the door to a greater possibility that such seafaring and the technologies and cultural philosophies may have also reached the new world Americas as suggested in this book. Further with the evidence presented of the correlations between Professor Thom's megalithic unit of measurements and the 27.5-inch spinal/light cubit presented here, this cultural travel and exchange could reach Egypt. For such a possibility I suggest not only their seafaring travels but the travels of Otzi the "Iceman" and jade axe heads.

A jade axe head was purportedly found near Stonehenge and dated to between 4000–3000 BC; similar jade axe heads have also been discovered in Scotland. Thankfully with today's science and technology, the jade/jadeite itself has been traced back to its origins high in the Italian Alps, easily 800 miles away in the mountains. The archeologists have clearly shown their origins and thus their journey. Where does Otzi come in? Otzi "the Iceman" was discovered in the Alps that border Italy and Austria at about a 10,000-foot elevation, frozen in ice. His body was discovered by tourists who thought a climber had died on the mountain. As recovery efforts were initiated, it was learned that this wasn't a modern climber; this was a climber dating back to 3300 BC who had died (possibly murdered) and had become an amazingly preserved, frozen, natural mummy. The knowledge that is being discovered from Otzi is an interesting window into life for him over 5,000 years ago, from health, tattoos, clothing, tools, weapons, and other paraphernalia. For the purposes here is the significance of his remains being found in the same area where the jade/jadeite axe heads found in Great Britain originated from. I wonder had Otzi already taken such jade and traveled to sites such as Stonehenge? Was he part of or did he trade with the megalithic seafaring culture? It has been shown such questions are well within the realm of possibility. It is not difficult to imagine that such travelers may have headed in the other direction, toward Egypt. The possibilities are tantalizing.

The archeological evidence and research have already been presented of an ancient European seafaring and megalith building culture, along with the trade and transport of axe heads from the Austrian-Italian Alps, including the remains of such a traveler found in those Alps. All this evidence traces back up to approximately 7,000 years ago. I also suggest that this also strengthens the theory and the evidence that an even earlier culture, going back over 20,000 years, may be the progenitor of the later megalithic seafarers. Further, these cultural ancestors may have been the group known as the Solutrean culture, a culture whose archeology has spawned a theory that this older Solutrean culture may have reached the Americas.

Welcome to the Solutrean hypothesis. This recent academic hypothesis presents the evidence and premise that North America eastern shores could have had humans arrive from Europe over 20,000 years ago! Perhaps for many readers who have been taught that humans came to North America about 11,000 years ago through the Bering Strait land bridge, this may not seem so controversial, particularly with the just shared evidence of a seafaring European culture of 7,000 years ago and that modern human beings have been around for over 200,000 years. That is not the case in today's archeology and anthropology; in those fields the Solutrean hypothesis is considered controversial. Suffice it to say that such ideas and change do not come easily. Not to get off track, but it may be of interest to read the history of the North America "Clovis First" hypothesis. Briefly, until the 1920s, archeologists believed there was not human habitation in the Americas until about 4,000 BC. In the late 1920s, tools, spear heads, and arrowheads began to be found, first in Clovis, New Mexico, and hence called Clovis points and Clovis culture. This moved habitation dates back to about 9,000 BC then 11,000 BC. As time went on archeologists began finding evidence of human habitation even earlier, 2,000 to 3,000 years earlier, especially in South America, but they were being ignored. It took an international council in 1999 to finally get agreement and move the Clovis line to about 13,000 BC. Even with that it is widely held and strongly defended that the Clovis were the first inhabitants of the Americas. It appears the Solutrean hypothesis is having a similar experience.

The Solutrean hypothesis has been put forth by the director of the Paleoindian/Paleoecology Program at the National Museum of Natural History at the Smithsonian Institution, anthropologist and archeologist Dr. Dennis Stanford, and University of Exeter archaeologist Bruce Bradley. Dr. Bradley and Dr. Stanford believe that the Solutrean culture that existed in what would be areas of modern France, Spain, and Portugal around 20000 BC crossed the Atlantic Ocean and arrived on the eastern shores of North America during that ancient time period. In doing so they were in North America thousands of years before the Clovis culture and that the Clovis culture was subsequent to the Solutrean arrival. Much of this premise is based upon stone and bone tools and weapon heads found along the East Coast of the United States and dating to the time of the Solutrean culture. Rather than describing what Doctors Stanford and Bradley have written, allow me to pay brief homage to their important work and honor the recent passing of Dr. Dennis Stanford by simply using excerpts from their academic paper.

> *So what are the implications for seeking a Clovis progenitor? It is our opinion that the well-developed Clovis bone and lithic technologies did not spring up overnight and there should be archaeological evidence of a transitional technology (p. 462).*

> **Solutrean is the only Old World archaeological culture that meets our criteria for an ancestral Clovis candidate. It is older than Clovis, its technology is amazingly similar to Clovis down to minute details of typology**

and manufacture technology, and the two cultures share many unique behaviours. Indeed, the degree of similarity is astounding (p. 462)

*We contend that the overwhelming and diverse number of similarities between Solutrean and Pre-Clovis/Clovis mediate against the simple explanation of convergence. The location, dating and technology represented by the Cactus Hill, Meadowcroft and Page-Ladson sites provide the "missing" chronological and technological links between Solutrean and Clovis. The hypothesis that a Solutrean Paleolithic maritime tradition ultimately gave rise to Clovis technology is supported by abundant archaeological evidence that would be considered conclusive were it not for the intervening ocean. None of the competing hypotheses for Clovis origins are supported by archaeological data. **Therefore, we argue that the hypothesis that a Solutrean Paleolithic maritime tradition gave rise to pre-Clovis and Clovis technologies should be elevated from moribund speculation to a highly viable research goal** (p. 473).[51]*

If this type of research is as fascinating to you as it is to me, I recommend *Across Atlantic Ice: The Origin of America's Clovis Culture* by Dennis J. Stanford and Bruce A. Bradley.

This chapter section on Stonehenge has brought us to a noteworthy point, first from the little known enigmatic builders of Stonehenge to the dramatic results that occur from using the spinal/light cubit measurement of 27.5 inches in its layout to Professor Thom's theory of the megalithic measurement system used to design such megalithic stone circles that collaborates with the spinal/light cubit. Then on to the recent research, evidence and theory of a 7,000-year-old European seafaring culture that was progenitor of Neolithic megalithic stone circles in Europe and Great Britain, including Stonehenge. It continues on to jade axe heads found in

Great Britain in the same time frame with this jade/jadeite originating from the Austrian/Italian Alps where the "world traveler" Otzi the Iceman mummy was discovered. Finally, on to the Solutrean hypothesis, which provides the research and evidence of a 20,000-year-old European culture who also appear to be seafarers, seafarers who made it to North America thousands of years earlier than most have surmised or are willing to accept.

Perhaps it has been noticed that two different academic archeological research studies have posited European seafaring cultures in their premises and indeed they do. What is intriguing and important is where they locate in Europe these seafaring cultures. Also linked are not only the stone origination of jade/jadeite axe heads, but where these finished axe heads have been discovered. They all overlap! Looking at the geographic maps of jadeite axe discoveries, the megalithic seafaring culture, and the Solutrean culture areas, one goes, "OMG." In particular the maps of Dr. Paulsson's seafaring, megalithic site-building culture and the areas of jadeite axe head discoveries overlap to such an extent that it provides prima facie evidence of links between them. As to the Solutrean culture some 10,000 years earlier than Dr. Paulsson's megalithic seafaring culture, finding evidence is incredibly difficult, but it is clearly there. Both cultures share geographic areas in France, Spain, and Portugal. I do not have the rights to present these maps here, but below I provide the links to them; a picture or map is worth a thousand words.

https://www.nms.ac.uk/explore-our-collections/stories/scottish-history-and-archaeology/stone-age-jade-from-the-alps/ (Jadeite axe heads)
https://www.pnas.org/content/116/9/3460 (Dr. Paulsson's megalithic culture)
https://en.wikipedia.org/wiki/Solutrean (Solutrean Culture)

Synopsis: Connecting the dots

This chapter focus has been a detailed presentation of the academic archeological research, evidence, and papers supporting the validation of the theory that there existed an ancient unit of measurement of 27.5 inches (called here a spirit/light cubit) used in the design and purpose of sacred sites across cultures and continents. Further that this proposed unit of measurement was derived and representative of the length of spine, not only as a body proportion, but also as a channel of spiritual higher consciousness (i.e., the spinal pathway of the kundalini serpent energy). Examples of such specific sacred sites were examined: the Great Pyramid in Egypt, the Kukulkan (El Castillo) in Mexico, the great kivas of the Ancestral Puebloans in North America, and Stonehenge in Great Britain.

In every one of the designated sacred sites, the evidence of a shared unit of measurement and its shared symbolism along with that specific culture's spiritual symbolism was significant. We started with the possible genesis of this spirit/light cubit in Egypt, where such a measurement is documented; the measurement results fit Ancient Egypt's Great Pyramid design, cultural, philosophy, and symbolism well empirically, down to the serpent and spine symbolism. Turning our attention to the other well-known pyramid-building culture of the Maya, the same can be written; the conclusion of Maya civilization academic researchers is that their engineers and architects used a unit of measurement equivalent to two 27.5-inch units as used in Egypt, approximating 55 inches, with evidence connecting to "serpent energy" with a focus toward creating an axis mundi, uniting Heaven and Earth. Then we moved on to the Ancestral Puebloan culture, which has clear links to Mesoamerican cultures, such as the Maya. The archeologists have posited, through analytical study of the Puebloan site of Paquime, in Casas Grandes, that their unit of measurement for construction design was 27.5 inches (70 cm), the same as the Egyptian measure and related to the Maya measure. There is also research of other Ancestral Puebloan sites having "great houses" and great kivas along with the traditional lore, which includes spinal symbolism and its importance in raising consciousness to connect to the Creator. Finally there is Stonehenge, where professor and professional engineer Alexander Thom provides us with possible units of measurement for megalithic stone circles sites which collaborate with the spirit/light cubit of 27.7 inches; a "megalithic rod" being equivalent to three spirit/light cubits. The conclusions of Professor Thom, which for so long have been highly controversial, are now getting support through a recent archeological paper that proposes Stonehenge and multiple other megalithic sites in Great Britain and Europe had their beginnings and technology from an ancient seafaring European culture. This supports both Professor Thom's theory of a megalithic system of measurements and my proposed theory of a further, more globally shared unit/system of measurement. This then leads to an even earlier European culture around 20000 BC (he Solutreans) who, evidence suggests, may have been the progenitor of the megalithic building culture. The Solutreans not only may have been the forerunners of the megalithic seafaring culture, the Solutreans themselves may have traveled to the eastern shores of North America, thus also being the forerunners of the Clovis culture. In that scenario they could have brought with them this proposed shared unit of measurement of 27.5 inches to the New World, which has shown to be evidenced both in Mesoamerican and North American Ancestral Puebloan cultures.

This may a good point to review and remind ourselves from Chapter 2 just how important a culture's units of measure are and the significance of finding shared units of measure between cultures.

The study of ancient measures used in a country is a basis of discovering the movements of civilization between countries. [52]

Measurement systems have provided the structure for addressing key concerns of cosmological belief systems, as well as the means for articulating relationships between human form, human action, and the world—*and new understanding of relationships between events in the terrestrial world and beyond.*[53]

Among the various tests of the mental capacity of man one of the most important ranking in modern life on an equality of with language is the appreciation of quantity, or notions of measurement and geometry. ...***Thus the possession of the same unit of measurement by different people implies either that it belonged to their common ancestors or else that a very powerful commercial intercourse has existed between them.***[54]

As I sit here and write this, even reviewing all the years of research and investigation accumulated in this journey providing the solid and compelling evidence for the theory that cultures on at least three different continents and some an ocean apart shared powerful "commercial" and spiritual connections thousands if not more than 10,000 years before most current theories believed possible, the implications are almost overwhelming.

The results presented here are archeologically and cultural philosophically paradigm changing.

This chapter section's purpose was to present the "hard" archeological evidence for an ancient multicultural, multi-continental unit of measurement that provided weighty evidence of a globally shared spiritual philosophy symbolized at sacred sites designed to connect Heaven and Earth and raise spiritual consciousness, presenting a concept of a shared human spirituality and cultural contact and communications thousands of years before such contact is considered possible. The evidence of contact through a shared unit of measurement is clear, and its spinal derivation and symbolism are plausible. What about for a shared spiritual philosophy and consciousness raising? Remember, all these sites were axis mundis and used for ceremonial and ritual purposes.

I waited to the end of the chapter to point out a major spiritual purpose implication that occurs from the studies concluding that the design and the construction of Stonehenge and similar sites by a European originating culture. Was it noticed in the archeological research in the Stonehenge section the reporting on two different papers about Stonehenge? One paper, specifically, showed that sound/sonic resonance could put the participants in either an alpha or deeper theta meditative state. The other research paper provided the evidence that a seafaring European culture was the originator and designer of megalithic structures throughout that area of Europe and Great Britain. Then, putting the conclusions of these two Stonehenge-related papers together, it presents that **these megalithic originators and inventors purposely designed such sites to spiritually raise consciousness**! Personally I think this is wonderful and a credit to our ancestors' pursuit of evolving consciousness. Archeologically and culturally I think it signals it is time to go back to Neolithic times and review the concepts of those cultures. Think of the implications of such a purposeful design; not only mind-expanding but mind-boggling.

Before moving on to the next chapters centering on the more spiritual and philosophical aspects of this journey, I would ask the reader to review this chapter and how "the dots connect," (a brief step away from the tapestry analogy).

The metaphor of connecting the dots seems appropriate. Humankind has been doing this since we could look up at the stars in the sky. Our consciousness has created wonderful beings and epic stories both physical and metaphysical by connecting those dots of light in the heavens. Some can say none of that is true, that it is just human imagination, yet since the existence of cultures, humans have looked up and connected the dots—seeing the past, the present, the future and the epic tales of gods and humans. It seems to be human nature. This dot-connecting nature applies both in the metaphysical and physical and can be a hybrid of both. These cultures not only connected the dots in the heavens; they connected them to Earth. In the physical this ability allows one to recognize patterns and their repetitions and predict events or probable conclusions. In any culture such observed repetition and predictability would itself be important and desired for survival and stability. I think this would be transferred to science and spirituality as much as possible. Amongst all the other specific evidence provided in this chapter, to connect the dots of this book's theory, humankind's nature displays that we are "wired" to do just that.

The evidence is compelling; the ramifications would be earth-shaking and paradigm-shifting to the archeological and anthropological communities. At the same time for the rest of humanity, I think it provides a hopeful and uplifting message left by these ancient cultures. The evidence demonstrates that these societies across the globe shared technology and a spiritual goal toward unity and Oneness through raising their consciousness beyond the physicality of Earth to a higher consciousness that united Heaven and Earth within themselves. This is a message that the world today could benefit from.

The chapters of this book to the point are the "hard" evidence for a culturally shared unit of measure: its physical body. The ensuing chapters will present its spirit.

> Mystic traveler, he's the unraveler
> And he will always bring you safely home
> Mystic traveler, he's the unraveler
> "Mystic Traveler" by Dave Mason 1977

5

The spine symbolized as a serpent—transitioning from the physical to the spiritual

My goal for the reader journeying to this stage has been to bring forth the archeologists' perspectives and archeological evidence that supports and reinforces the premise that multiple ancient cultures on multiple continents shared a common unit of measurement in the layout and plans of their sacred sites; that this measurement was codified from the length of the human spine. I believe this proposition is amply supported with the evidence put forth in the prior chapters. Such sites were created to symbolize axis mundis, places where Heaven and Earth unite. Additionally the premise puts forth that such sites were considered focal points to raise human consciousness above finite physicality toward a spiritual universal consciousness, a Oneness with all. This is a journey of both the physical and the spirit. The physical evidence for such a measurement is strong, straightforward, captivating, and, frankly, elegant. It is both plausible and possible. Though this journey for me has taken years, decades even, in many ways the measurement unit itself is the least complicated, the least convoluted aspect of this research and these correlations. I think the greatest opposition will be that, despite the evidence that points toward a sharing and communication between widely separated cultures, it is far before today's archeology deems it was possible. Though, aptly enough, if one looks at the historic timeline of these current theories, they have changed dramatically over time. (Remember the increasing evidence and support of the Solutrean hypothesis or the appearance dates of modern Homo sapiens, now moved back more than 200,000 years ago, or for that matter the multiple megaliths site carbon dated to at least 9600 BC [Gobekli Tepe in Turkey] as three examples). The shared measurement and spiritual philosophy of raising consciousness, our core being, exhibits a humanity-wide effort and purpose to bring Heaven to Earth—are these cultural memories of a utopia, a golden time? I wonder.

It is the greater, higher symbolic purpose of this measurement unit based on the length of the spine and why its use is shown so prominently at sacred sites that takes one into a more rarified sphere of the consciousness of metaphysics and spiritual philosophy. As noted in Chapter 3, there is a clear challenge of interpreting the thought behind prehistoric art; that same challenge is present in interpreting more recent ancient spiritual philosophies. Hopefully, as my shoes on this journey were "re-souled," so are yours. As previously quoted, archeologists and scholars of ancient metrology tell us that evidence of a shared unit of measurement is an indicator of a shared original culture or active trade and communication between cultures. These experts further state that measurements exhibit a thought process beyond the surrounding physical world, providing links between the material and immaterial and relationships between events in the terrestrial world and beyond.

The intent of this chapter is to continue beyond the earlier chapters that have built the physical evidence and connections, moving from the body to the spirit into that elevated and diversified sphere of spiritual philosophies.

The premise and the evidence presented are that these ancient cultures created a measurement that in physicality could frame space and in spirituality and consciousness could frame the infinite.

I noted in Chapter 1 that multiple ancient cultures understood the physical connection and avenue of the spine to the brain, with this physical avenue providing a pathway for higher spiritual consciousness. Where the discussed unit of measure represented the length of the spine, the visual presentation of the spine is quite often symbolized by the serpent. My personal realization of this was a cross between an "aha" and an "oh my gosh how did I not see this before" moment. This is a perfect example of such symbolism being hidden in plain sight. The visual presentation of the spine representing a serpent is impossible not to see, as evidenced in the following images.

Fig. 28 Human spine   **Fig. 29** Serpent

I would hazard to say, as a reader, the response has been the same as mine; the spine makes an incredible depiction of a serpent.

As this chapter begins I think we must initially address the general historic negative assertions that serpent symbolism has in the

Judeo-Christian tradition. Though serpent symbolism throughout the world is predominately positive, in Judeo-Christian beliefs, serpent symbolism seems quite the reverse, being interpreted predominately as harmful.

I researched deeper into the Bible and found there are many positive, constructive stories using serpent symbolism that I had not been taught. It was so clarifying and enlightening in the spiritual journey. It was another epiphany "hidden in plain sight" that will be shared in more detail later in this chapter. First, though, let's continue with the cultures that have been explored here and address their use of serpent symbolism.

The image of the serpent, sometimes also depicted as a dragon, is one of the oldest and widespread symbols in mythology and spiritual philosophies. It is a symbol for both spiritual and physical fruitfulness. In all these spiritual beliefs, the serpent was seen generally as positive. That view did not mean that caution should not be exercised also; serpent symbolism for the spiritual was both a sign post and a warning for the journey of expanding one's consciousness. These cultures recognized the positive energy of the serpent but also warned of the cautionary aspects. This approach to "serpent energy" can be considered much like the production and use of electricity, a vital and integral aspect to the advancement of civilization, yet if improperly used or not understood, it can do great harm. I think if you substitute the word spirituality for electricity in the last sentence, most would agree with the comparison. I imagine at some point we all have been illuminated and shocked by either or both.

The Hindu and Buddhist spiritual philosophies are very straightforward with this symbolism connecting the spine and the serpent; there it is represented through the spiritual concept toward raising consciousness called the kundalini. The term kundalini is a Sanskrit word loosely translated as "coiled serpent." Basically, it is the pathway through the spine and central nervous system of the subtle (energetic/spiritual) body, passing through the six chakras (spiritual focal points in the physical body) ultimately to the seventh (crown) chakra at the top of the head. It is by following this pathway, one reaches enlightenment; universal consciousness, uniting the physical and the spiritual. This sounds very familiar to the Hopi perspective shared in Chapter 4, does it not?

The principle of this book is this very journey going beyond the Eastern philosophy to a more global, cross-cultural journey and intuitive desire of raising our consciousness and embracing the spiritual in the physical. As others have phrased it, we are not physical beings having a spiritual experience; rather we are spiritual/consciousness beings having a physical experience.

The possibility of this physical global-wide spiritual journey has been shown as conceivable. Awakening, elevating, and properly channeling this spiritual, subtle body toward greater consciousness beyond the physical has a rich history in many cultures—as noted, being more obvious in some than others. The Hindu and Buddhist traditions reinforce such purpose with the tale of Buddha gaining enlightenment under the Bodhi tree, and his protection by Mucalinda, the king of the serpent gods (Nagas). Mucalinda wrapped the Buddha in seven coils of his body and rose above to shield him with a seven-headed cobra hood while Buddha reached enlightenment. This symbolic serpent path to higher consciousness will be explored with this chapter "yoking" them, primarily, with the four cultures that Chapter 4 showed shared a common unit of measurement derived from the length of spine used in the design of their sacred sites.

The spine symbolized as a serpent 105

Fig. 30 Buddha reaching enlightenment protected by Mucalinda

A. Egypt

The ancient Egyptians had the Uraeus (a symbol of the goddess Wadjet), a cobra emanating from the forehead of their crowns while some representations showed it with wings. Wadjet, just as the Egyptian serpent deity Nehebu-Kau, was one of the oldest, pre-dynastic gods of Egypt. The ancient Egyptian lore shares how the Wadjet serpent emanating from the forehead came to be. In one version the Egyptian god Ra (also known as Atum) sent his daughter, Wadjet, to be his eye and find two of his other children who had become lost. When Wadjet returned with them he was so overjoyed he rewarded Wadjet by placing her on his forehead in the form of an upright cobra to always be close to him and protect him. Wadjet was given the title of "the eye of Ra." This title was also held by other goddesses, including Isis. In another version Ra plucked out his eye and sent it to find his two lost children. When his eye returned with the lost siblings, Ra had grown a new eye; his original eye felt betrayed, so to reward and to conciliate his original eye, Ra changed it into a cobra and placed it on his forehead.

It is not farfetched to observe from both these versions of the creation of the Uraeus that Ra developed a "third" eye (the Eye of Ra) in the middle of his forehead. For many, this may be familiar from the Hindu and Buddhist traditions of raising the kundalini energy up through the seven charkas with the sixth chakra being known as the "third eye" or Ajan chakra. Chakras are considered subtle energy focal points linked to the physical body generally in a vertical line following the spine. The Ajan chakra represents the spiritual eye of the subtle, spiritual body and is located in the center of the forehead between and just above the eyebrows. The similarities between the Eastern and the Egyptian concept is self-evident, both using the symbolism of the eye and the serpent.

The spine symbolized as a serpent 107

Fig. 31 Uraeus: Abu Simbel Temple: Egypt

Fig. 32 Winged Wadjet serpent

There are other links to the Egyptian "Eye of Ra." Some scholars believe that the sun disc that is encircled by two Uraeus serpents also symbolized Ra's eye. The god Ra also wore as his symbol the disc of the sun encircled by the serpent Khut. Khut was a representation of the goddess Isis, and in this form she was the "light giver," the "light of the New Year," As a further note the Great Pyramid is reported to have also been known as "Ta Khut"; the light/the flame.

Fig. 33 Eye of Ra Kings Valley: Egypt

Fig. 34 Khut Sun Disk Kings Valley: Egypt

Along with serpent symbolism, light and spiritual light concepts were important aspects of Egyptian philosophy. This is not only seen in the serpent wrapped sun disk symbols often found on or above their heads, but the stone shafts of their obelisks represented frozen rays of light and their pyramids also depicted frozen light as stairways to Heaven. The pyramid themselves were considered "cosmic engines" to unite the soul's ka and ba to become an akh or aakhu; a radiant being of light. Remember the Egyptian serpent Deity Nehebu-Kau, who aided the Egyptian soul in uniting the ka and ba. This also bears a similarity to the Eastern philosophy of raising the kundalini light energy to reach enlightenment, the luminosity or clear light of a Buddha-consciousness.

In Chapter 4's Egypt section, Nehebu-Kau was touched on lightly in the relation of having the same root word, nb, as a documented Egyptian unit of length of 27.5 inches called a nbi. The root word transliterates as to unite or to yoke together. This 27.5-inch unit of measure I identify in this book as a spirit/light cubit derived from the length of the spine. I believe there is a correlation with the nbi unit of measure identified as 27.5 inches to the enigmatic Egyptian unit of measure called an "aakhu meh." I believe that these measures are the same length. This conclusion comes from the names of both measurements and the links to the purpose of Nehebu-Kau and the Great Pyramid; nbi is related to uniting the ka and ba of the soul in completing the soul journey to become an aakhu. The related names of these measurements show a "during" and "completed' process of that same journey. Akin to the hieroglyphic and the hieratic writings of ancient Egyptians, hieroglyphs are sacred writings; hieratic is priestly writing, similar but having a different hierarchy of power and importance, one more to the spiritual (aakhu meh) and one more to the physical (nbi).

Nehebu-Kau or "He Who Unites the ka" was a benevolent snake god whom the Egyptians believed was one of the original primeval gods. He assisted in fulfilling the Egyptians' ultimate purpose of uniting the ka and the ba, aspects of the Egyptian soul that basically represented the original divine spark or spirit, part of the soul and one's individual part of the soul respectively. In uniting one's ka and ba, one would become an aakhu (akh or khu), a radiant being of light living eternally in the northern sky, where the stars never set (circumpolar stars). Nehebu-Kau was a god of protection who watched over the pharaoh and all Egyptians, both in life and the afterlife. It was believed he gained his power after swallowing seven cobras, and he was the one who gave each Egyptian his true name and fed him with the milk of light. He was on occasion depicted with a head at each end of the body. Nehebu-Kau was linked to the sun god, swimming around in the primeval waters before creation, and then bound to the sun god since time began. In one version when he received his power by swallowing seven cobras, his power was so much that Ra the sun god would have to put his nail in his spine to control him, which is interesting considering the presented symbolism of the spine and spiritual energy. In today's practices of raising consciousness through meditation and other practices, raising the kundalini can also happened unintentionally or unexpectedly and is called a "kundalini syndrome," which can overwhelm the individual. Perhaps the Egyptian tale is a warning.

Nehebu-Kau's purpose, as a spiritual being, is the same as the "cosmic engine" physical presence of the Great Pyramid: to unite the soul's ka and ba in an aakhu (radiant being of life) fulfilling the individual journey and purpose in life. Then there are other similarities such as the mysterious unit of measurement noted earlier, the "aakhu meh." We know that "aakhu" means radiant being of light or spirit of light, and "meh" generally follows a name of an Egyptian measure; it can mean a binding, more specifically a headband crown binding. This presents the possibility of a depiction much like the Uraeus, the serpent emanating from the pharaoh's forehead, balancing spiritual power and control, perhaps symbolizing the prevention of kundalini syndrome. Meh can represent binding an area of measurement or, in this case, spiritual energy. Ancient Egyptian spiritual philosophy is intricate.

Fig. 35 Nehebu-Kau with two heads; Kings Valley Egypt

Then there is the curious tale of Nehebu-Kau swallowing seven cobras for his power, showing similarities to the Eastern philosophy of seven chakras (spiritual centers in the body). This Eastern similarity continues with the tradition of Buddha reaching spiritual enlightenment while being protected by the serpent deity king Mucalinda with his seven heads and seven coils. There are also the different hieroglyphic writings for the Egyptian word "mer," which translates as *pyramid* but also as *sacred serpent*. This, too, fits well in the theme.

The importance and symbolism of the serpent in Egypt's spiritual philosophy are overwhelmingly pervasive and included in its greatest remaining monument standing today, the Great Pyramid, not just in the proposed unit of measurement and its symbolism. I am referring to the British anthropologist and the first director of the England and Ireland Archeological Society. C. Staniland Wake states it succinctly: "The Great Pyramid is thus a monument not only of Sabaism, but of serpent worship..."[55] (Author note: Sabaism is the worship of the spirits in stars as part of the host of heaven.) From the Egyptians' serpent deity's purpose to the associated proof presented here, C. Staniland Wake does seem to put it concisely.

B. The Maya

This segues us to the next pyramid builders who also had serpent symbolism as an important core aspect to their spiritual philosophy and were also amazing stargazers/astronomers: the Maya. It has already been shared that investigators have concluded that the Mayan unit of measurement in their engineering and architecture was a measure of approximately 55 inches (a zapal), a double unit of the 27.5-inch spirit/light Egyptian measurement. One aspect for validating the Maya measurement was using it to measure ten different structures in three different ancient Maya cities. Provided earlier was a sample of their results for the Kukulkan Pyramid, called El Castillo from the Spanish, located in the Maya city of Chichen Itza, Mexico.

The Kukulkan Pyramid was dedicated to the Maya god of the same name, meaning "plumed/feathered serpent." The Maya pyramids are different from their Egyptian counterparts as the tops are truncated, with a flat top upon which sits a temple room. Maya pyramids were known as places of ceremony and ritual. Many such pyramid temples can be found through the Maya empire dedicated to the plumed serpent and this deity's variants. Variants considered closely related to Kukulkan include the deity Q'uq'umatz, the feathered serpent of the Popol Vuh (the book of the Quiche Maya people, located more in Guatemala), and Quetzalcoatl, the feathered serpent deity of the Aztecs. In each case they are depicted as feathered/winged serpents. This is not a low-lying serpent in the grass; this is a winged serpent that has raised its energy upward. Being a winged/feathered serpent, it combines the aspect of an Earth-related serpent and the heavens-related bird, thus linking Heaven and Earth together. I see this comparable to the winged and/or upright serpent symbolism in Egypt accentuating the positive force of spiritual energy just as has been described in Eastern traditions: the kundalini. Scant information of these related serpent deities exists. Unfortunately, due to the burning of almost all the Maya books by Spanish priests, only four Maya codices (books) are known to have survived, and what is considered a complete translation of the Maya hieroglyphic writing has only been recently achieved.

In general these were creation deities with links between Heaven and Earth; they were intermingled with the Maya vision serpent who connected the physical world with the spiritual world. The importance of serpent symbolism was on a similar level to the Egyptian civilization. The importance of Maya serpent symbolism, spiritual energy, and connecting Heaven and Earth can be vividly seen during the equinoxes at the Kukulkan Pyramid. This is another example of an axis mundi sacred site. What occurs is a profound and dramatic creation of a Serpent of Light from the temple at the top to the base of the north stairs, where the carved serpent head becomes illuminated. (See picture below.)

Just think of the advanced astronomy, engineering, and architectural design expertise that went into creating this pyramid, coordinating it with the heavens for this event. Disappointingly, though I have been to Chichen Itza, I have not had the opportunity to be there to experience this event. I would suggest to the reader, as I have, to go on YouTube and search for this event. Even on video it is an incredible sight to watch this serpent of light being created. It brilliantly shows the prominence of serpent symbolism for the Maya. A serpent of light! If you look carefully, you'll see the serpent's body of light consists of seven triangular coils, bringing us back to the significance of the number seven, serpents, and light itself cited earlier pertaining to Egypt and Eastern enlightenment. There is even

Fig. 36 Kukulkan Pyramid equinox Serpent of Light. Note the 7 coils

Fig. 37 Kukulkan Pyramid north staircase base serpent heads

evidence of a possible earlier name of Chichen Itza, meaning seven great rulers or seven bushy places—more indications of a connection to the cross-cultural prominence of the number seven. Another interesting fact is that, though Maya feathered/plumed serpents are invariably risen or upright, here at this temple the serpent's head is at the base on the ground. This apparent juxtaposition makes sense when viewed as representing spiritual energy rising up through the spine. Look back toward the beginning of this chapter at the image of the spine next to a serpent; the head would be at the base of the spine, and the rising energy to consciousness at the serpent's tail. This would seem to answer such a reversal. All this, including the serpent symbolizing a link to Heaven and Earth, this "light serpent," and the Maya zapal measurement (55 inches) used here, now leads to the Maya ceremonial double serpent bar.

Fig. 38 Maya king statue holding the double serpent bar: Copan, Honduras

The Maya ceremonial double serpent bar is an essential and important iconographic symbol for the Maya philosophy. As shown in the prior picture, it is often seen being held by a Maya king as a status symbol and link to heavenly power. It combines the symbolism of the sky and serpent; it is considered that the Maya purposely used the same-sounding words, khan (sky) and kaan (serpent), in such fashion and was also connected to the Maya vision serpent and a heavenly vision journey. This also sounds similar to the Egyptian word mer, which can mean pyramid and sacred serpent using the same word with different hieroglyphs.[56]

Simply put by the experts:

"…the ceremonial bar identified its holder as a personification of the 'world tree' or **the axis mundi thereby directly connecting the holder to cosmic deities and the celestial realm.**

The ceremonial bar was consistently associated with primordial energy, with shamanic performance and transformation…"[57] (Author's emphasis.)

Fig. 39 Maya ceremonial double serpent bar: Copan, Honduras

The importance of this serpent symbol is self-evident: connecting the holder with both Heaven and Earth, raising the holder's spiritual energy, and transforming him. This continues to fit the premise of this book—the sacred sites and the raising of the spiritual serpent energy toward divine enlightenment. It also is appropriate to be embodied in a double serpent and in doing so smoothly transitions into the designs of their temple, such as the Kukulkan Pyramid. It can be seen that the Maya zapal (55 inches and representing two spinal lengths) and the Maya double (two) serpent bar both mirror symbols of uniting Heaven and Earth while raising one's consciousness. Truly, the design of the Kukulkan Pyramid, the symbolism of the double serpent bar, and the integration of the zapal measurement are elegant.

C. Ancestral Puebloans

As the journey continues tracing the spiritual symbolism of the serpent and spine, we follow the same path of Chapter 4's physical evidence onward to the Ancestral Puebloans. The spiritual philosophy of the ancient Puebloans is well hidden in the mists of time. What has been deduced is primarily through the intrepid work of Southwest archeologists of the pottery, architecture, petroglyphs, and traditions of their descendants such as the Hopi and Zuni and what they are willing to share.

Perhaps one of the earliest seeds of information is in the tale of how the Hopi were able to leave the lower earlier worlds into the present fourth world. As reviewed in Chapter 4 they were able to rise up into the higher world through the growth of a giant hollow reed reaching up into the sky into the next world. By climbing up through the reed's hollow core, they were able to arrive in the higher world through the opening at the top of the reed. This reed opening to the higher world is symbolized in the floor of the kivas of the Puebloans and is called the sipapuni. The Puebloans were able to receive the seed for this giant reed through their entreaties to Baholinkonga.[58] Baholinkonga is a plumed serpent; some alternatives show him as a horned serpent. Horn symbolism, like the serpent, was also normally related to metaphysical powers. Baholinkonga provides a strong connection to the Maya and Aztec plumed serpent symbolism. This symbolism also continues the connection of raising the spiritual (serpent) energies up through the spine, with Baholinkonga supplying the seed for this reed that leads them up to a higher world.

Fig. 40 Upright horned serpent petroglyph. Albuquerque, NM

The continued importance of serpent symbolism can be observed in the Puebloans descendants, the Hopis, through their spiritual ceremonies exampled in the annual snake dance, properly called the Snake-Antelope ceremony. Only the Hopis and a select few chosen by them have ever seen the entire ceremony, much of it conducted in the sacred kivas. The public used to be allowed to attend the outdoor courtyard snake dance stage itself, but due to the poor manners of the non-Hopi, this opportunity is no more. Imagine people coming to anyone's religious service, eating, chatting, and even laughing during it, and one can understand why.

The Snake-Antelope ceremony on its surface is pragmatic to bring needed rain from the heavens for the crops in the earth, but like most spiritual ceremonies, it has multiple levels; it is also a spiritual marriage uniting Heaven and Earth at the level of all creation, raising and renewing nature and humans themselves, a rebirth and resurrection. This can be surmised by the nature of the ceremony needing both the Hopi Snake society and Antelope society; by combining both the snake and the antelope, the serpent energy is raised. The symbolic earthly energy of the serpent with the horns coming from the top or crown of the head of the antelope, the top of the head for the Hopi, as explained in Chapter 4 is where they connect with God. This serpent-antelope comes together as a metaphysical horned serpent akin to Baholinkonga.

Fig. 41 Lightning serpents sand painting altar: artist depiction

120   The Spirit of Light Cubit

Fig. 42 Lightning serpents petroglyph, Valley of Fire State Park, Nevada

This symbolism of a risen serpent and its energy can be seen in a rare picture of a sacred altar corn/sand painting (ca. 1898) inside the antelope kiva at the pueblo of Walpi, Arizona. My focus of this sacred painting is its symbolism of the four zigzag figures depicting "lightning snakes," which have triangular heads each with a horn. I believe they not only represent lightning occurring during a thunderstorm or a downpour of rain, but also the metaphysical union of Heaven and Earth through spiritual serpent energy such as the kundalini. The photograph next to it is a petroglyph from Valley of Fire State Park, Nevada, considered to represent such lightning serpents also. Please notice the vertical "ladder" image next to it. Petroglyph interpretations note this as a possible emergence ladder of the kiva or the hollow reed ladder to emerge in the upper world. This would seem appropriate with the lightning serpents and the concept of raising spiritual consciousness.

Along the same lines of these just described pictures is an image that can be found inside and up on the wall of a tower structure in the Mesa Verde, Colorado, located at the Cliff Palace cliff dwellings.

Fig. 43 Interior wall painting, Cliff Palace Tower

Fig. 44 Exterior of Cliff Palace Tower

On the right is the tower itself and on the left is the painting that is inside the upper wall of the tower. If you are flexible enough and bend over backward through the lower opening, you can see it high up on the wall. Fortunately at the time of this opportunity and photograph, I was. It can be seen from comparison with the previous images that these appear to be lightning (energy) snakes also, four on a side divided by what possibly may be, and according to petroglyph interpretations, a single pole ladder or a stalk of corn. In the multiple levels of the snake/antelope ceremony, both would be appropriate, bringing rain for the corn (maize) and raising the spiritual energy. As for the ladder/cornstalk separating the two groups of four lightning serpents, it certainly can remind one of a spine, as could the earlier petroglyph reed/ladder.

Finally, below this framed image in the tower, one can see pyramid shape images on each side.

Fig. 45 Same wall painting wider view

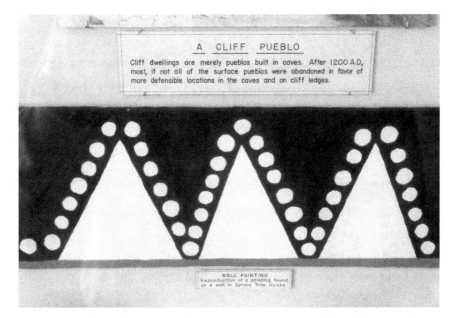

Fig. 46 Spruce Tree House wall painting depiction

In a wider photo you can see that below the lightning serpent tableau are painted two sets of three triangular shapes; next to it I have put a picture of a reproduction that can be found in the Mesa Verde National Park museum of a painting found on a wall of Spruce Tree House, another cliff dwelling located in Mesa Verde National Park. It is thought that they may represent mountains. I know some will consider this wild speculation—and it is speculation, just not wild. I find it very curious such groups of triangular shapes images are depicted at Puebloan sites with the evidence put forth here of a shared unit of measurement that has its earliest discovered origins in Egypt. I leave it there and will now retreat back across the ocean to Stonehenge.

D. Stonehenge

The only visual direct connection, if it can be called that, to serpent symbolism at Stonehenge is originally in the early 1700s with respected British antiquarian (archeologist) William Stukeley, who was leading the belief that Stonehenge was a Druid temple and incorporated their serpent symbolism. He further believed this symbolism could be traced back to Egypt. Dr. Stukeley is credited with promoting and advancing academic investigation and using such methodologies of accurate measurements and documentation of archeological sites, much like the later archeologist Sir Flinders Petrie was credited for in the archeology of Egypt. Stukeley was certain not only that Stonehenge was a Druid temple but that the Avebury (Abury) stone circles were a similar Druid temple. Stonehenge may be the most recognized stone circle in the world, but the Avebury circle is the largest megalithic stone circle in the world. It is believed that its construction was circa 3000 BC, versus the stone circle aspects of Stonehenge being placed circa 2400 BC.

Stukeley's estimated the dates of construction of both Stonehenge and Avebury are over 2,000 years later than the current archeological dating with its advantage of more advanced dating technology. Due to the current archeological construction dates, his Druid temple theory has been discarded, primarily because the Celtic culture, with their Druid priests, is determined to have arrived from the European Gaul area, as called by the Romans, sometime between 1000 and 600 BC. These arrival dates are much later than the construction of the stone circles, thus delegitimizing Stukeley's theory.

As displayed in Chapter 4, with the new research and evidence pertaining to these megalithic structures, Stukeley's theory can no longer be just swept away out of hand. There I point to the academic report that shows the evidence and the research that the British Isles megaliths' construction and design originated from a seafaring culture dated about to 5000 BC, located in Europe, including western France, in the same area of Celtic Gaul. This area in Western France contains Brittany and borders the coast along the Atlantic Ocean and the English Channel. I had similarly noted earlier this is an area which through the millennia has had a line of cultures from the Solutreans, circa 20000 BC, through the megalithic seafaring culture, circa 5000. This area's habitation continues to the Celtic/ Gaul culture and their Druids, circa 1000 BC, on to the Roman Empire, circa 100 BC, continuing in a curious kind of reverse migration of ancient Celts from Great Briton, circa 400 AD back to this pivotal area of megalithic building. This timeline and this region's focal point allow for the distinct possibility of cross-cultural seeding and sharing of serpent symbolism in this area's ancient spiritual philosophies. The historical documentation of the prevalence and importance of serpent symbolism worldwide would aid in such synergistic interaction, including the Celtic priests, the Druids.

This discourse brings us back full circle, like the serpent holding its own tail, a symbol of the Ouroboros, to Dr. Stukeley and the megalithic serpent temples of Stonehenge and Avebury. The Celtic culture and their Druids may have come later than the construction of Stonehenge and Avebury, but they could well have brought

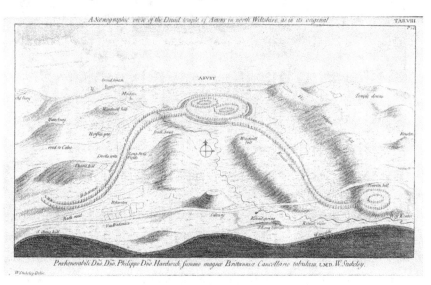

Fig. 47 William Stukeley's artist depiction of the original possible Avebury stones

with them the knowledge of earlier cultures and serpent symbolism and could have known the significance and purposes of these stone circles. Below is Stukeley's drawing for the Avebury megalithic stone circles and colonnades, imagining the original condition of the site. Admittedly Dr. Stukeley took some artistic license to create his finished vision.

Stukeley's work clearly inspired Great Britain's incredible mystic, artist, and poet, William Blake, in his work "Jerusalem: The Emanation of the Giant Albion." This was beautifully created by Blake in the poem's last plate.

Fig. 48 Jerusalem: The Emanation of the Giant Albion, p. 100, PD-1923

Blake's plate combines Avebury and Stonehenge in his depiction of the serpent temple. The interpretation of the symbolism of Blake's final plate of his epic work is mixed, being seen as possibly negative or positive. Since serpent symbolism has both aspects, negative as a low serpent in the grass or positive as an upright or winged serpent, this seems apt. Blake's plate has been seen as the serpent representing the repression of the spirit, but it also has been interpreted as positive, representing "the new consciousness of eternity in time rather than beyond it."[59] Admittedly I smiled at this interpretation as I saw and have shared here that the serpent symbolism and its representation in the 27.5-inch unit of length makes an ideal pendulum timepiece, thus creating the same concept of time and space, Heaven and Earth united. This concept appears to be captured in Blake's poem "Europe a Prophecy":

"Thought chang'd the Infinite to a Serpent..." Then was the Serpent temple form'd, image of Infinite, Shut up in finite revolutions..." (Author's emphasis.)

Blake's ambiguity of serpent symbolism can also be seen in this poem. In the full poem the symbolism is toward the negative aspects—our infinite spiritual selves having been trapped in the finite, creating a separation from unity and Oneness—but hope can also be seen in the recognition of our greater non-physical self that waits within to be released. This message is also suggested in "Jerusalem." I take hope and comfort in that the research of the use of Stonehenge shared in Chapter 4 points to its purpose for raising and reconnecting our higher selves by raising consciousness through meditation.

Beyond William Blake's metaphysical poems and art and William Stukeley's depiction of Avebury with the archeological evidence that the Celtic Druids could well have cross-cultural communications with the descendants of the earlier megalith builders, the Celtics' ship building and seafaring may be evidence of this. The Celtic seafaring people, also known as the Veneti, of around 400 BC inhabited Brittany, France, the time of the Druids, the same area of the 5000 BC shipbuilders' megalithic culture and the earlier, 20000 BC, Solutreans.

Strabo, a Greek historian circa 20 BC, described the Veneti Celtic ships as solidly designed, oak-built, with leather sails to handle the unforgiving sea conditions and force of the North-Atlantic winds. Author Kevin Duffy writes, "…their timber sailing vessels which have been described up to 150 feet long with a beam of 16 feet. The *Mayflower* which carried the Pilgrims from England to America in 1620 was only 90 feet long."[60] Please note: For even more perspective of their ship building ability, Christopher Columbus' 1492 expedition ships were: Santa Maria: 160 feet long with a 18-foot beam; Pinta: 56 feet long with a 16-foot beam; Nina: 50 feet long with a 16-foot beam—and they all made it to America from Spain. The *Golden Hind*, which Sir Francis Drake sailed for his famous circumnavigation of the globe in 1577, was 102 feet long with a beam of 20 feet. The first circumnavigation of the globe was completed by the Ferdinand Magellan expedition in 1522 with the ship *Victoria*, whose length has been estimated at a length of 65 feet. As not to digress further into the ancient seafaring capabilities, which have been shown to be highly capable, let me return to the Druids' interpretation and use of serpent symbolism.

The Druids of ancient France (Gaul) and Great Britain were the priestly group who were the knowledge- and wisdom-keepers, all being passed on orally as they had no known writing system. They were the doctors, teachers, advisors, judges, and spiritual leaders of the society. The use of serpent symbolism was part of their philosophy; one example is the adder stone or serpent stone. The serpent stone is somewhat geode-like with a naturally occurring hole through it. The lore on them has them created by the solidified saliva of a group of serpents or from the head of a serpent. The serpent stones value was for healing both in the stone's presence and from a distance.

More important and central to the Druid spiritual philosophy than the healing serpent stones is the Druid deity HU, whose symbol was the serpent, as explained by Manly P. Hall:

The spine symbolized as a serpent   129

Fig. 49 Serpent Stone

Their temples wherein the sacred fire was preserved were generally situated on eminences and in dense groves of oak, and assumed various forms—circular, because a circle was the emblem of the universe; oval, in allusion to the mundane egg, from which issued, according to the traditions of many nations, the universe, or, according to others, our first parents; serpentine, because **a serpent was the symbol of Hu** (Author's emphasis: this will become more significant in a later chapter), *the Druidic Osiris; cruciform, because a cross is an emblem of regeneration; or winged, to represent the motion of the Divine Spirit.*[61]

Here not only is the importance of serpent symbolism shown but also, on occasion, it's being represented with wings "to represent the motion of Divine Spirit," bringing us back to the winged/plumed/

upright symbolism of risen spiritual consciousness demonstrated for the Egyptians, Maya, and Ancestral Puebloan cultures who also share the same unit of measurement and serpent symbolism focused upon in this work. The lore of early Briton shows links to such narratives of the Druids and serpent symbolism with the King Arthur legends.

Uther Pendragon was the father of Arthur Pendragon, with the magical help of Merlin. Merlin becomes mentor and counselor to Arthur, who becomes hero and king to the burgeoning kingdom of Briton. Merlin is generally believed to be a bard and a Druid, while both Uther and Arthur's name Pendragon translates as chief or head dragon. The dragon is another version of the symbol of a winged serpent. The Druid and serpent symbolism can still be seen in the legends of Briton. Interestingly as time went on and Christianity became prominent in the British Isles, the King Arthur legends shifted to a much more Christian theme, i.e., the Holy Grail quest. Another example of the entrance and influence of Christianity into the British Isles is the tale of St. Patrick, the Christian missionary and bishop of Ireland driving out the serpents. Perhaps this is symbolic of the rise of Christianity pushing out Druidism.

E. Christianity

The example of the Christian St. Patrick driving out the serpents (Druids) from Ireland is a good shift to address a question I had about serpent symbolism, and perhaps the reader has also. Though Christianity's sacred sites and spiritual philosophy are not part of any of the evidence for premises and conclusions that have been proposed here, many are probably aware of the negative connotations of serpent symbolism in Christianity. As noted in the beginning of the chapter, though serpent symbolism throughout the world is predominately positive, with the caveat of properly handling serpents or serpent (spiritual) energy, in Christianity, serpent symbolism is predominately negative.

I grew up in the Christian faith, and like many others, I was taught early and often that serpents were bad; serpents were evil. That if it wasn't for that evil serpent in the Garden of Eden, humanity wouldn't have fallen and we would all still be there. Through my childhood and into my adult life, my perception of serpents and their symbolism was that they were trouble and we should stay away from them. Then as I began studying and reading about other religions, ancient mystery schools, and mythology, I discovered that the serpent symbolism was far from one-dimensional, that it was much richer, diverse, and positive. My question to myself was; what was going on with Christianity and the serpent versus so many other spiritual philosophies? So I delved deeper into the Bible and found there are many positive, constructive stories using serpent symbolism. More accurately the evil aspect of serpent symbolism that was drilled into me from a young age was so prominent that the positive examples did not register in my consciousness. Following are several positive examples:

Exodus 4 (King James Version) 2. And the LORD said unto him, what is that in thine hand? And he said a rod. 3. And he said, Cast it on the ground. And he cast it on the ground, and it became a serpent; and Moses fled from before it. 4. And the LORD said unto Moses, Put forth thine hand, and take it by the tail. And he put forth his hand, and caught it, and it became a rod in his hand:

Here God is speaking to Moses to go to the Pharaoh and tell him to let the Hebrews go. Moses tells God Pharaoh would not believe or obey Moses on his word alone, so God gives Moses some signs to show that he speaks for God. One is having Moses' staff being able to turn into a serpent. I did have to smile a bit at this passage when God turns Moses' staff into a serpent; Moses runs away, a very human response even if you are speaking with God. God has to tell him to pick (raise) it up. The point is that this is a positive symbolic use of the serpent to be a sign and symbol to Pharaoh that Moses' words come from God.

> *Exodus 7 (King James Version) 9 When Pharaoh shall speak unto you, saying, Shew a miracle for you: then thou shalt say unto Aaron, Take thy rod, and cast it before Pharaoh, and it shall become a serpent. 10 And Moses and Aaron went in unto Pharaoh, and they did so as the LORD had commanded: and Aaron cast down his rod before Pharaoh, and before his servants, and it became a serpent. 11 Then Pharaoh also called the wise men and the sorcerers: now the magicians of Egypt, they also did in like manner with their enchantments. 12 For they cast down every man his rod, and they became serpents: but Aaron's rod swallowed up their rods.*

This example has interesting aspects; it is not Moses' staff but Aaron's staff that turns into a serpent showing they were both messengers from God. The passage also shows the Egyptians were familiar with such symbolism and serpent energy that their "wise men and sorcerers" came and turned their staff into serpents. Consider the staffs representing spines and the serpents symbolizing the raising of spiritual energy. Aaron and Moses had a higher spiritual consciousness, being closer to God, and then consuming the lower energy of the Egyptians.

> *Numbers 21 (King James Version) 6 And the LORD sent fiery serpents among the people, and they bit the people; and much people of Israel died… 8 And the LORD said unto Moses, Make thee a fiery serpent, and set it upon a pole: and it shall come to pass, that every one that is bitten, when he looketh upon it, shall live.*

The Hebrews in their long exodus began losing their spiritual way, experiencing the challenges and dangers of lower-self spiritual energy not risen. Moses is directed by God to raise a serpent on a pole, and by looking up, raising the serpent energy into higher consciousness, they will be saved.

John 3:14-21 (King James Version) 14 And as Moses lifted up the serpent in the wilderness, even so must the Son of man be lifted up: 15 That whosoever believeth in him should not perish, but have eternal life.

In the New Testament the same serpent energy symbolism can be seen and the importance that the Son of man must raise up the spiritual energy.

Matthew 10:16 King James Version (KJV) 16 Behold, I send you forth as sheep in the midst of wolves: be ye therefore wise as serpents, and harmless as doves.

This is a very telling example of higher serpent energy. Jesus is talking to and advising his disciples how to spread the spiritual message, with the wisdom of the serpent and the gentleness and peace of the dove. The serpent and the bird are being combined, creating the winged/plumed serpent, the symbol that has been shown throughout this book to represent raising the spiritual energy into higher consciousness; universal consciousness. The symbolism is clear and compelling.

Even the Hebrew translation of "seraph"—one of the orders of angels, celestial, radiant beings of light, messengers of God—literally translates as "burning ones" and is considered synonymous in the Hebrew bible for serpent.

All these examples are positive representations of serpent symbolism in Christianity. Finally I'll present an example from the Bible that shows spinal symbolism.

Genesis 3 (King James Version) 13 And the LORD God said unto the woman, What is this that thou hast done? And the woman said, The serpent beguiled me, and I did eat… 15 And I will put enmity between thee and the woman, and between thy seed and her seed; it shall bruise thy head, and thou shalt bruise his heel.

This brings us back to where the overwhelming negative serpent symbolism is birthed from. I used the term "birthed from" because it is appropriate for the symbolism that can be seen here. This verse describes the birthing of a child, a birthing that will bruise the serpent's head and bruise the child's heel. This can be seen symbolically describing the physiological effects that occur during childbirth. In preparation and during birth, the body releases hormones to allow the stretching of the pelvic joints and its ligaments (sacroiliac) to aid in an easier birth. The pelvic girdle, which encompasses all of this, has its center with the sacrum. The sacrum is the fused spinal vertebrae S1 through S5, and its name is traced back to the Greeks, meaning "sacred bone." The sacrum is the symbolic serpent's head depicted by the spine (please refer back to the image of the spine shown at the beginning of this chapter). The pelvic girdle ligaments with the sacrum during birth are stretched out and widened, essentially bruised in the act of birth. Many women express how their hips have widened from childbirth. Then as the child is born, in a normal delivery, it is a cephalic or a head-first delivery. In such a normal delivery, as the feet appear last, it is not uncommon for the newborn's extremities, particularly the lower extremities, to have a slight bluish hue, like a bruise, prior to the baby's first breaths oxygenating the body.

The comparison to the biblical description here is not difficult to see. Eve's seed; the child, the serpent's seed; the coiled serpent/spiritual energy at the base of the spine to the bruising of the serpent's head; the stretching of the ligaments surrounding the sacrum, culminating in the child's bruised heel; the lower extremity bluish hue from being unoxygenated. Seeing the symbolism is not a hard stretch, nowhere near the stretch of the pelvic girdle a woman endures in childbirth. I thank you all for what you endure.

These examples clearly show that there are many positive connotations to serpent symbolism, along with warnings. In Christianity, unfortunately, the most press has gone to the negative side. Serpent symbolism throughout the ages and the world, including Christianity, as shown, is related to positive spiritual energy with warnings of accessing through our consciousness such powerful energy. What all of these spiritual beliefs warn us is that it is the lower self, the snake in the grass, which causes the trouble and not the risen, upright, or winged serpent representing our higher spiritual self. They all share the idea of raising our consciousness for a higher spiritual benevolent purpose to guide us back to an Eden-like consciousness.

> We are stardust
> We are golden
> And we've got to get ourselves
> Back to the garden
> "Woodstock" Joni Mitchell 1970

6

What does this mean for ancient societies? What happened?

I was going to write how this journey began with a neglected unit of measurement in Egypt, but that wouldn't be true. The reader knows from Chapter 1 this journey began much earlier for me, but even then I was still asking the same questions: what does it all mean, what happened, what's the message? Only then it was about my life, our journeys, and our purpose, the apparent separation of the physical and spiritual with our consciousness somewhere in between. As I wandered and searched along on that journey, the roads unexpectedly merged with this ancient unit of measurement. If this was a movie, the screenwriter would probably be suspected of creating a deus ex machina plot device to resolve my journey's search. I assure you it is not.

Even as I first came across a mention of a 27.5-inch measurement used in temples and pyramids in ancient Egypt, I thought it was curious in its mystery and worthy of exploring, but I had no idea that what I thought was an interesting side path on my journey would end up being the road that not only answered my questions about this unit of measurement but brought answers and clarity to many of my questions of life.

Before I continue further on that path, let me briefly review the premises of this book and the evidence supporting them. I do not think it can be reiterated enough.

Beginning with the study of ancient cultures' units of measurement and their significance and importance:

"The study of ancient measures used in a country is a basis of discovering the movements of civilization between countries."[62]

"Among the various tests of the mental capacity of man one of the most important, ranking in modern life on an equality of with language is the appreciation of quantity, or notions of measurement and geometry. …Thus the possession of the same unit of measurement by different people implies either that it belonged to their common ancestors or else that a very powerful commercial intercourse has existed between them."[63]

"Cultures sharing similar measurement systems likely had some form of contact. Should such a measurement be located in architectural remains, and appear in halves or doubles, then the probability that this measure reflects a real historical unit of measure increases."[64]

I could have used other quotes from the few more modern scholars of ancient metrology, but almost to a person in their own papers they cite and quote Petrie. In this arena scholars still recognize, like Petrie, that a sign of a culture's thought process advancement is a system of metrology and that if a unit of measurement or its halves or multiples is discovered among cultures, it is indicative of shared ancestors or strong communications between the cultures.

Here is such a shared unit of measurement among cultures, starting in Egypt:

- Ancient Egypt has a documented unit of measure of 27.5 inches (70 cm). Its use also provides significant numerical results.
- Ancestral Puebloans of North America have a documented unit of measure of 27.5 inches (70 cm). Its use also provides significant numerical results.
- The Maya have a documented unit of measure of approximately 55 inches (140 cm) which is 2 x 27.5 inches. Its use also provides significant numerical results.
- The builders of Stonehenge in the British Isles have research evidencing a possible use in the megalith of a measurement unit of 81.7 inches (207.5 cm) equivalent to three 27.5-inch measures (82.5 inches, 210 cm). Its use also provides significant numerical results.
- Though some of these cultures are separated by appreciable time and distance, they all meet the criteria of a shared unit of measurement reinforced with significant numerical results at each sacred site.
- Body proportions: the origins of ancient measurements. The length of the human spine as proposed for the origination of this measurement has been demonstrated as feasible and significant to sacred sites and serpent symbolism. The recognition of the spine in particular for Egypt and Ancestral Puebloans is clear. The serpent symbolism in the cultures of the Egyptians, the Maya, and the Ancestral Puebloans is pervasive, and the described possible connection to Stonehenge with Britain and European Celts and their Druid priests having their serpent symbolism is also persuasive.
- Cultural philosophies uniting time and space: 27.5 inches is an ideal measure for a pendulum. It was shown that this length makes an ideal pendulum for timekeeping, thus creating a measurement that incorporates not only physicality

but time. All of these sites are cultural axis mundis, and this measurement with its symbolism and use accentuates the concept of uniting Heaven and Earth at these sites.

This is a very brief synopsis of what has been already described in depth; returning to what it means, the ramifications are staggering. The current dates of the four sacred sites' construction periods investigated here with the shared unit of measure are in an approximate range from 2400 BC for both the Great Pyramid and Stonehenge (for the Stonehenge megalithic stones), 1000 AD for the Ancestral Puebloans, and 850 AD for the Maya Kukulkan Pyramid. These sites have a time period range of 3,400 years plus a geographic range of three continents and the Atlantic Ocean.

I would conjecture from all that has been presented that most would at least accept the possibility of this shared unit of measurement and, hence, communications between cultures, between ancient Egypt and ancient Britain. The recent findings, evidence, and conclusions of an archeologist concerning a 5000 BC seafaring megalithic culture located in Brittany, France, and Egypt's own coastline along the Mediterranean Sea makes communication a comfortable likelihood. This likelihood is increased with the discovery of Neolithic mummy "Otzi," dated to before 3,000 BC—hundreds of years before the Great Pyramid and Stonehenge are believed to have been constructed. Otzi was found in the Italian Alps in the same area scientists have determined Neolithic jadeite for ceremonial axe heads found near Stonehenge and up into Scotland were mined. If such distances could be sailed and climbed northwest from the Alps to Britain, there is little reason that such geography could not be traveled southeast to Egypt. It is actually a pretty straight line, point to point.

I would further conjecture that this premise and its evidence for shared unit of measurement(s) with the Maya to the Ancestral

Puebloan would be accepted particularly with the recent determination of the same measure being used in Paquime, Casas Grandes, Mexico; after all, it is already known there was trade and communication between them, and contact would be a purely land route endeavor.

The resistance will be about some type of connection or communication over the Atlantic Ocean between these cultures. To accept or even entertain what archeologists who study ancient metrology propose, that shared measures indicate communications or a communal ancestral group, from this evidence and the premise would indicate such communication across the Atlantic occurred, at least thousands of years earlier than presently believed. Such possibilities and proposals, rightfully so, are not easily considered. The ocean—ah, there's the rub...to wake to the possibility and perchance to dream of such connections.

Fortunately this proposal and evidence have not been left in dry dock. There are archeologists who, from different research and evidence, have proposed that the Atlantic was crossed not just thousands of years ago, but over 15,000 years ago.

To travel back to these many millennia, let us first return to Chapter 4 in the Stonehenge section. Here is archeologist Dr. B. Schulz Paulsson and her work presenting the premise that megalithic structures—such as stone circles, including Stonehenge in Great Britain and the area of Europe emanating out from Brittany, France—had their genesis from a seafaring megalithic building society dating to 5,000 BC. As noted in her own words:

"The megalithic movements must have been powerful to spread with such rapidity at the different phases, and the maritime skills, knowledge, and technology of these societies must have been much more developed than hitherto presumed. This prompts a radical reassessment of the megalithic horizons and invites the opening of a new scientific debate regarding the maritime mobility and organization of Neolithic societies, the nature of these interactions through time, and the rise of seafaring."[65]

And we saw earlier her results are already being supported:

Dr. Michael Parker Pearson, an archeologist and Stonehenge expert supports the conclusions. "This demonstrates absolutely that Brittany is the origin of the European megalithic phenomenon."[66]

Dr. Paulsson's statements that I felt were particularly powerful pertained to the Neolithic society's organization, maritime skills, knowledge, and technology being much more advanced than presently recognized and their seafaring mobility allowed greater and more distant interactions with other societies. These statements along with the analysis and study of megalithic builders mirrors the same conclusions that scholars of ancient metrology arrive at when finding different cultures and societies using the same or similar units of measurements; that these cultures at least had shared some kind of communication.

The evidence and conclusions which include the possible shared unit of measurement, the travel of jadeite from the Italian Alps to the plains and hills of Briton, the trekking of Otzi and seafaring of the European megalithic builders move the possibility of contact and communication from ancient Briton to ancient Egypt to a probability.

The above described 5000 BC proposed cultural exchanges from Egypt to Briton is the first leg of this journey through space and time. The next leg travels back an additional 10,000 to 15,000 years (20000 BC – 15000 BC) and across what presently is 3,700 miles of open ocean. Archeologist Dr. Stanford points out that the time period in question is during the last Ice Age into the Younger Dryas period with the ice cap moving down significantly into the Northern Atlantic, creating an artificial coastline that could be navigated along with the hunting of seals and seabirds similar to the Inuit, an indigenous people of the Arctic regions. Interestingly and possibly very telling of this much earlier and longer journey is that it starts in the very same area of the described megalithic building seafarers' culture of 5000 BC. This earlier culture is known as the Solutreans.

Also in Chapter 4 was shared the Solutrean hypothesis overview of Dr. Stanford and Dr. Bradley, whose evidentiary premise is this "stone age" Solutrean culture traveled by ocean to the East Coast of North America. I had shared their overall evidence of a "pre-Clovis" culture, meaning cultures prior to approximately 11500 BC; the Solutrean culture is dated circa 20000 to 15000 BC. Briefly here as a refresher:

Bone and stone tools have been discovered at five different sites along the East Coast of North America at sites dated from 16000 to 21000 BC. These tools correspond to the same unique style of tools as the Solutrean culture's tools discovered in Europe. These tools are not just generally similar as Dr. Stanford points out: "We can match each one of 18 styles up to the sites in Europe."[67]

As it has been pointed out, it is the ocean that continues to be the rub, no matter the evidence, whether it is tools or a shared unit of measurement. As Dr. Bradley and Dr. Stanford conclude:

> *The hypothesis that a Solutrean Palaeolithic maritime tradition ultimately gave rise to Clovis technology is supported by abundant archaeological evidence that would be considered conclusive were it not for the intervening ocean. (p. 473)*

> *None of the competing hypotheses for Clovis origins are supported by archaeological data. Therefore, we argue that the hypothesis that a Solutrean Palaeolithic maritime tradition gave rise to pre-Clovis and Clovis technologies should be elevated from moribund speculation to a highly viable research goal. (p. 473)*[68]

As I noted earlier, I believe the validity of their conclusion is aided with the recent evidence of a 7,000-year-old European seafaring megalithic culture in the same area where the Solutrean society resided, perhaps their descendants. The validity of their conclusion is strengthened by the evidence presented here of a shared unit of measurement unit occurring on both sides of the Atlantic supported by scholars of ancient metrology that such measures are evidence of communication between them. As the Solutreans are proposed to be the progenitors of the Clovis culture through the evidence of shared tool technology found and dated in the ancient strata they were buried in, I propose such an ancient culture crossing the Atlantic by the evidence of a shared unit of measure used by the descendant cultures of North America in their sacred structure sites with the measure's progenitor found in Europe and Egypt. Two separate approaches based on two different lines of evidence arriving at the same conclusion only increases the possibility of its reality. Such a reality would rewrite human history: again this is staggering. The possibility of a society such as the Solutreans some 20,000 years ago navigating the globe and by doing so distributing their technology and measurements leads to what I believe go hand in hand in such contact and technological distribution: the distribution of their spiritual philosophy. If the known history of multiple civilizations traveling to new lands is any indicator, they invariably brought their religions/spiritual philosophies to share or convert others with them. In this case I propose a spiritual philosophy incorporating serpent symbolism was shared also.

It should also be considered that a culture comparatively more advanced technologically than other cultures of the time period would probably be viewed in an elevated status not only technologically but also in spiritual philosophy. In the latter scenario the culture having had such contact may convert or absorb the serpent symbolism to their own spiritual philosophy; if they already had such symbolism, they may merge theirs into the culture they may see in higher esteem and consciousness.

The serpent is considered one of the oldest and most globally spread spiritual symbols found. This symbolism is positive and powerful and carries warnings of negative results if such power is misused. The previous chapter went into detail of these aspects in cultures, specifically the cultures who shared this described unit of measure. It could be argued that serpent symbolism is so global that it is a human consciousness archetype that could have occurred through all these cultures independently. Even if serpent symbolism in different cultures occurred independently, it would then aid in creating spiritual bonds in contacts with foreign cultures. An example of this would be the Spanish conquistador Hernan Cortes making contact with the Mesoamerican culture of the Aztecs and their ruler Montezuma II. Though the veracity is still debated today, Cortes records that the Aztecs thought they (the Spanish) were gods returning and Cortes was the manifestation of Quetzalcoatl, their plumed serpent deity. Unfortunately, this was not the case, nor was a sharing of positive serpent symbolism, as the Spanish and Christianity at the time saw serpent symbolism as evil. Christianity is an outlier to most cultures' view of serpent symbolism.

The almost universal primordial use and significance of the serpent in ancient spiritual philosophies is also an indication of early humans and their societies having a more advanced thought process (consciousness) that included the immaterial, the spiritual concepts as discussed in Chapter 3. There has been a recent discovery in Botswana, located near the southern tip of Africa, that helps confirm this symbolism and consciousness, moving these possibilities back to around 70,000 BC. In brief:

This summer, inside a cave in remote hills in the Kalahari Desert of Botswana, archaeologists discovered evidence of organized ritual activity dating to 70,000 years ago. Their findings are set to change our perceptions of human development since previously it was generally thought that human intelligence had not evolved the capacity to perform group rituals until perhaps 40,000 years ago, when the first such evidence appears in Europe.

The Botswanan evidence comprises a stone carving resembling, so the team believes, a snake 2m high and 6m long. "You could see the mouth and eyes of the snake. The play of sunlight over the indentations gave them the appearance of snake skin. At night, the firelight gave one the feeling that the snake was actually moving," said Associate Professor Sheila Coulson, from the University of Oslo, who along with Dr. Nick Walker has been conducting research in northwest Botswana for 8 years as members of the University of Botswana/Tromso Collaborative Programme for San Research.[69]

This finding, with its implications toward serpent symbolism and earlier evolved human intelligence, certainly increases the possibility of the existence of a culture not only later continuing with a more evolved philosophy with serpent symbolism but also with more evolved technological abilities to navigate the seas and build megalithic structures. After all, they would have another 50,000 years to develop both.

This incredibly ancient discovery in Botswana of serpent symbolism and the prevalent ancient symbolism of the serpent found all over the world speaks to the depths of the human consciousness and a resonance that humanity as a whole seems to share from the most ancient times. Psychologist Carl Jung, the founder of analytical psychology, addresses this in his concept of the collective unconscious:

"The existence of the collective unconscious means that individual consciousness is anything but a tabula rasa and is not immune to predetermining influences. On the contrary, it is in the highest degree influenced by inherited presuppositions, quite apart from the unavoidable influences exerted upon it by the environment. **The collective unconscious comprises in itself the psychic life of our ancestors' right back to the earliest beginnings.**"[70] (Author's emphasis.)

Now here is where I move into the challenging, rarified realm of interpreting human consciousness of this proposed ancient society. Though I believe my premise has a valid base, I tread lightly. First, I will reiterate an earlier quote on such matters.

"Measurement systems have provided the structure for addressing key concerns of cosmological belief systems, as well as the means for articulating relationships between human form, human action, and the world—**and new understanding of relationships between events in the terrestrial world and beyond."**[71] (Author's emphasis.)

The measurement, the spirit/light cubit, and all its aspects that have been expounded upon here provide "the structure for addressing key concerns of cosmological belief systems" quite elegantly. Additionally the spirit/light cubit premise put forth further offers a "new understanding of relationships between events in the terrestrial world and beyond." Yes, it is clearly symbolic, representing the spine and serpent energy at axis mundi sacred sites to unite Heaven and Earth in the participants.

In my view the presented archeological and anthropological evidence points to a time period at least 15,000 years ago for the possibility of an advanced technological civilization capable of global sea navigation and megalithic type construction for sacred sites. This unknown, lost civilization, perhaps now called the Solutreans, along with their technology brought with them a unit of measure I identify as the spirit/light cubit. This measurement was used for the architectural and engineering designs. Please note as I define their advanced technology, think at least the technology we are familiar with from the 18th century, the 1700s, and remember the earlier citation of the presently known earliest circumnavigation, which occurred in 1522.

This lost civilization also brought with them an evolving higher consciousness spiritual philosophy using serpent symbolism. They had united their technology and spirituality into their units of measure with the prime key unit of measurement being a 27.5-inch measure that represented the spine and the kundalini serpent-type subtle energy that rises through it to expand consciousness both physically and spiritually. (See Chapter 1.)

The historic records show that ancient cultures throughout the world were familiar with and incorporated into their own spiritual philosophies a positive metaphysical symbolism of the serpent. A symbolism that evidence suggests is so archetypal in human consciousness, it seems to go back at least 70,000 years in humans' thought processes. This shared symbolism could have been the sign for acceptance and cooperation between otherwise disparate cultures as contact was made.

With this acceptance and cooperation, rather than the Cortes example, like the earlier cited seafaring megalithic builders of Dr. Paulsson, this lost civilization helped create around the world sacred sites using this measurement. These sacred sites represented axis mundis, places where Heaven and Earth came together not only through marking heavenly events (solstices and equinoxes, for example) but to do the same in human consciousness, in spirit. These sites were designed to assist and synergize the raising of consciousness through meditative trancelike states. The evidence for this purpose has already been provided in this book for the four sacred sites exampled.

To avoid such a purpose being dismissed out of hand, let me do an abridged review of the practice of meditation. In the beginning of this journey, back in Chapter 1, I wrote of the importance of meditation and its long history in spiritual philosophies around the world for reaching enlightenment. Science labels these results as behavioral changes. This practice is found under many different names and techniques, but the purpose is the same: raising awareness and spiritual consciousness. Presently the term "mindfulness" is being used instead of meditation, a rose by any other name. You would be amazed just by putting the word "meditation" in an Internet search how prevalent, ancient, and global meditation was and is. Also noted in Chapter 1 were, as science and technology "catch up" to meditation practices, the physical effects that are being discovered, from changes in one's EEG to actual physical changes and growth of the brain itself. The latest neuroscience research is even providing evidence that meditation enhances altruism between each other. I highly suggest the reader look into the research on meditation being done by Harvard professor and PhD Sara Lazar; it is amazing and thought-provoking.

Just imagine, with all this evidence and background, there existed an ancient advanced civilization navigating the globe, connecting with other cultures, sharing their technology and spiritual philosophy. This lost civilization assisting all these cultures in building sacred sites to assist in raising their consciousness, evolving their brains, and making us more altruistic and united with each other; efforts toward a worldwide Utopia, a true community of humankind. This is the other part of my premise for the ancient past and my hope for the future.

This may be an odd tangent, but this proposition makes me think of Stanley Kubrick's movie *2001: A Space Odyssey*, in particular the enigmatic black monolith scene in the beginning of the movie, where you see a prehistoric Earth and a group of hominids, early ancestors of humans. These hominids wake up to find this large black monolith by their rocky recess. First they are scared and agitated by its presence; then as they calm down and approach it, they all begin to touch it in a kind of awe. The last shot of that scene is directly up from the monolith at a rising sun above the monolith with a sickle moon above the sun. The next scene cuts to the hominid group scavenging for food, and one of them picks up a bone and begins to use it as a tool and then a weapon. This movie clip was to show, from this mysterious monolith, the evolution of thought process toward modern human beings.

What I find fascinating are the core similarities of this scene of a monolith/megalith evolving hominid consciousness to the premise of megalithic type sites created to evolve the consciousness of modern humans. Unfortunately, in my view, it shows human evolution in a violent weapon-using fashion. I wonder if these scenes in the movie are a result of a kind of traumatic collective consciousness memory caused from some disastrous event that so shattered the proposed burgeoning utopian society labors that these efforts became an anathema in human consciousness.

This leads to:

What happened?

In the human consciousness and cultures from ancient times, there comes to us many stories of lost paradises, lost civilizations, and the fall of golden ages. They can be found in humankind's myths, lore, and religions. For lost civilizations I will use only one example, Plato's writing about Atlantis in his Timaeus and Critias writings from around 360 BC. Atlantis is a quintessential example not only of an ancient lost civilization, but one that had a golden age, an Eden-like existence, then a dramatic fall from grace and destruction around 9600 BC. I have with, great brevity, reduced Plato's Atlantis saga to excerpts relevant to the subject under consideration. The first two excerpts pertain to what happened to Atlantis, what researchers' scientific evidence today has led them to theorize about a cataclysmic event which may have occurred circa 10,900 BC to 9600 BC. The second excerpt eerily continues in this prescient fashion relating what would occur after such a cataclysmic event; the severe regression of impacted cultures and the loss of their records and knowledge. (I have bolded some parts of these excerpts.)

"You are young, the old priest replied, young in soul (Greek; psyche meaning soul, mind or spirit), every one of you. Your souls lack any learning made hoary by time. The reason for that is this: **There have been and will continue to be numerous disasters that have destroyed human life in many kinds of ways.** The most serious of those involve fire and water, while the lesser ones have numerous other causes. And so also among your people the tale is told that Phaeton, child of the sun, once harnessed his father's chariot, but was unable to drive it along his father's course. He ended up burning everything on earth's surface and was destroyed himself when a lightning bolt struck him. **This tale is told as a myth, but the truth behind it is that there is a deviation in the heavenly bodies that travel around the earth, which causes huge fires that destroy what is on the earth across vast stretches of time.**"[72] Note; this is an Egyptian priest speaking, which I find curious with the earliest genesis of the light/spirit measurement being from Egypt.

"In your case, on the other hand, as in that of others, **no sooner have you achieved literacy and all the other resources that cities require,** than there again, after the usual number of years, **comes the heavenly flood. It sweeps upon you like a plague, and leaves only you illiterate and uncultured people behind.**"[73]

The next and final two excerpts from Plato described not the physical destruction of Atlantis but rather the spiritual and moral demise of a society that had been reaching heights of a higher consciousness of ethics, peace, and unity, losing its spiritual way and then falling from those great heights.

"They possessed conceptions that were true and

entirely lofty. And in their attitude to the disasters and chance events that constantly befall men and in their relations with one another **they exhibited a combination of mildness and prudence, because, except for virtue, they held all else in disdain** and thought of their present good fortune of no consequence. They bore their vast wealth of gold and other possessions without difficulty, treating them as if they were a burden. **They did not become intoxicated with the luxury** of the life their wealth made possible; they did not lose their self-control and slip into decline, **but in their sober judgment they could see distinctly that even their very wealth increased with their amity and its companion, virtue. But they saw that both wealth and concord decline as possessions become pursued and honored. And virtue perishes with them as well.**

"Now, because these were their thoughts and because of their divine nature that survived in them, they prospered greatly as we have already related. **But when the divine portion in them began to grow faint as it was often blended with great checkers of morality and as their human nature gradually gained ascendancy,** at that moment, in their inability to bear their great good fortune, **they became disordered**. To whoever had eyes to see they appeared hideous, since they were losing the finest of what were once their most treasured possessions. But to those who **were blind to the true way of life oriented to happiness** it was at this time that they gave the semblance of being beauteous and blessed. Yet inwardly they were filled with an unjust lust for possessions and power. But as Zeus, god of gods, reigning as king according to law, could clearly see this state of affairs, **he observed this noble race lying in this abject state and resolved to punish them and to make them more careful and harmonious as a result of their chastisement.**"[74]

Plato's narratives some 2,500 years ago are a fascinating

commentary on his own civilization and any eras' civilizations toward a tendency of hubris that they are the most advanced and the most knowledgeable civilization about the world and all its events. Plato's narrative posits that the rise and fall of civilizations and cataclysmic events go further back in time and have a longer history than is realized. That these are cyclical events that go beyond recorded history. Plato even explains what contributes to our lack of such recorded history, not only the great lengths of time that history disappears into but what catastrophic events do to civilizations. These events basically would revert such civilizations back from their current state to such a point that in the struggle for basic survival, their science and records are lost. In modern warfare the phrase "bomb them into the Stone Age," though a sad testimony on war, brings out the point of what catastrophic events could do to a civilization. Such civilization-shattering events and the massive trauma they cause could take thousands of years to recover from. Think about the kind of memories and tales that would survive; I used the analogy earlier of a game of "telephone," passing a message through thousands of people over thousands of years; what would be left of the original message? Plato also notes that such events, over such a long timeline to recover from, could well be remembered as fables woven into mythology or stories of their religions. Plato gives us a specific example of such an occurrence of a solar or possibly a comet or meteor causing such a world disaster: "a deviation in the heavenly bodies that travel around the earth, which causes huge fires that destroy what is on the earth across vast stretches of time." He then tells how this became a tale in mythology.

Plato further shows that there is a second phase that weaves such a natural cataclysmic event into the mythos of why the event occurs. This is presented in the second pair of quotes from Plato in reference to the fall and destruction of Atlantis. Plato lays out how the noble Atlanteans from their golden age began to lose their "divine portion" of higher consciousness, that linking of Heaven and Earth within themselves, and as they lost this higher altruistic mind-set, they lost themselves into a more carnal physical consciousness. The god, Zeus, saw this fall and determined to punish them, a punishment to stop their fall from getting worse or going to a point of no return to the divine. This also provides an explanation of how survivors of such a cataclysmic heavenly event would explain, metaphysically, why such catastrophe would be visited upon them; that the leading civilization of the time displeased god(s) in hubris and/or debased earthly actions, devolving themselves rather than evolving themselves.

Plato's tale certainly has echoes of the Garden of Eden from the Bible. In Chapter 5, I indicated that serpent symbolism in Christianity was not all negative, that this symbolism had positive connotations also that serpent symbolism spiritual energy had the caveat that improperly used or raised was negative and dangerous. I would postulate that one of the meanings of Adam and Eve and the serpent, in this perfect garden (world), is that this spiritual energy was improperly used causing their (humankinds') fall. I think about the verse where after they improperly used this energy, they fell more into their lower physical selves, into a lower consciousness where they felt naked and hid. I think this interpretation may be reinforced by another verse where God is looking for them and asks, "Where are you?" God could find them physically if their consciousness was in physicality as of yet. His question was about where their consciousness was. This idea is reinforced by another verse in KJV Genesis 3 after this: "Unto Adam also and to his wife did the LORD God make coats of skins, and clothed them." God clothed them in "coats of skins," and their consciousness became more physical. Ultimately they were driven out of this perfect world westward by an angel (a celestial being) with a flaming sword into a harsh physical world.

Another biblical example I feel also echoes Plato's tale and observations would be the Tower of Babel. Again, I bold some sections.

Genesis 11 King James Version (KJV):

> **11 And the whole earth was of one language, and of one speech.** 2 And it came to pass, as they journeyed from the east, that they found a plain in the land of Shinar; and they dwelt there. 3 And they said one to another, Go to, let us make brick, and burn them thoroughly. And they had brick for stone, and slime had they for morter. 4 And they said, Go to, **let us build us a city and a tower, whose top may reach unto heaven**; and let us make us a name, lest we be scattered abroad upon the face of the whole earth. 5 And the Lord came down to see the city and the tower, which the children of men builded. 6 **And the Lord said, Behold, the people is one, and they have all one language; and this they begin to do: and now nothing will be restrained from them, which they have imagined to do. 7 Go to, let us go down, and there confound their language, that they may not understand one another's speech. 8 So the Lord scattered them abroad from thence upon the face of all the earth: and they left off to build the city.** 9 Therefore is the name of it called Babel; because the Lord did there confound the language of all the earth: and from thence did the Lord scatter them abroad upon the face of all the earth.

The synopsis of the Tower of Babel narrative depicts a society migrating westward (from the east) across the world, able to communicate with each other. They decide to build a tower to reach heaven to become like gods themselves. I want to refer back to Chapter 1, with its discussion of apotheosis; the chapter notes, biblically or otherwise, the concept of apotheosis, to become divine, godly, is not a negative aspiration in its pure form; if anything I showed it is our ultimate journey, becoming friends of God. This journey is safest and most directly accomplished through spiritual types of meditative practices and applications to raise such energy in ourselves, to reunite Heaven and Earth. That is what I believe this narrative is also a lesson of—it is not accomplished through the physical (i.e., building a tower to reach heaven) but the spiritual, building a consciousness to reach heaven. Yes, I write of physical structures being designed and built around the world, but these structures were not designed to physically reach heaven. They were to assist one in consciousness to connect to or unite with Heaven. The following will give an explanation of how such narratives of heavenly reaching sites and serpent symbolism became skewed as the cause of divine retribution coming down from the heavens.

This biblical narrative is often seen as a creation story of how the multiple cultures and languages of the world occurred. I think this narrative along with the biblical narrative of the Garden of Eden and Plato's narrative of the rise and fall of Atlantis can also be seen through the lens of the evidence and premise of this book. The premise being that there was an advanced ancient society probably over 15,000 years ago that had both global seafaring and megalithic-type construction abilities. This ancient society also had a spiritual philosophy based on serpent symbolism that represented, through the spine, the kundalini type of spiritual/consciousness energy, and that they brought this more evolved and applied serpent symbolism purpose to cultures to assist in evolving consciousness. History shows that cultures probably already incorporated positive serpent symbolism in their philosophies, which would assist in the acceptance of such an application.

This brings us closer toward the "what happened." What happened to this evidentiary-based, proposed ancient global seafaring, megalithic designing, higher consciousness society sharing their technology and spiritual philosophy? The evidence for the possibility of such a society shared here is ample and compelling; is there evidence of a cataclysmic event to send it "back to the Stone Age"? The answer is yes.

There are two such evidenced global civilization-shattering events whose proposals will be presented; one is the theory of a CME (Coronal Mass Ejection), where such an ejection from our sun struck the earth, and the other is that a comet(s) or meteor(s) impacted the earth, the Younger Dryas Event (YDE).

In the theory of a CME striking the earth, a main proponent of the theory is Dr. Robert Schoch, a geologist and geophysicist who for years has been researching and writing quite ably on ancient civilizations and their mysteries. He bases his conclusions and applies his background and scientific methods on the work of plasma physicist Dr. Anthony Peratt and his team's work identifying prehistoric petroglyphs around the world that are apparently visual recordings in stone of powerful plasma events in the atmosphere. Such events could be caused by a CME; less powerful events such as solar flares create those beautiful atmospheric events known as the aurora borealis. At the level of a CME, the beautiful atmospheric events become destructive.

Dr. Schoch, incorporating Dr. Peratt's research with his own, presents the theory that such a devastating CME hit the earth around 9700 BC. Dr. Schoch describes the results of such a plasma event (a CME); its high energy levels could fuse rock, burn any and all flammable materials, vaporize bodies of water, melt glaciers, and create a domino effect toward even more global environmental disasters. He continues that such an event would eliminate any advanced civilization of the time and perhaps may even impact their mental abilities. As I had noted Plato's narratives of memories of ancient Atlantis and its destruction, Dr. Schoch also notes his theory aligns with such a 9700 BC event. I would suggest reading his well-written book on this subject published in 2012, *Forgotten Civilization*.

Adding veracity and evidence for such a possible CME event and ironically occurring in 2012 was a "near miss" of a massive CME just passing by Earth. This immense CME shot out from the sun on July 23, 2012; if it had occurred a week earlier, it would have made a direct hit on Earth. In a NASA article on this CME near miss two years after the event, Dr. Daniel Baker of the University of Colorado states, "If it had hit, we would still be picking up the pieces."[75] The article further states: "Analysts believe that a direct hit by an extreme CME such as the one that missed Earth in July 2012 could cause widespread power blackouts, disabling everything that plugs into a wall socket. Most people wouldn't even be able to flush their toilet because urban water supplies largely rely on electric pumps." The article continues that the odds have been calculated of such an event striking Earth is 12 percent in the next ten years!

Pause to think about what is being described here; a massive global loss of electrical power impacting everything from water, food, refrigeration, transportation, communication, and medical care. Then there is the domino effect of such disruption and chaos would cause to populations, with long-term ramifications. The article states that such an event would be destructive enough "to knock modern civilization back to the 18th century." That is the 1700s, before electricity, at time of horse-drawn carriages and sailing ships. This is a very sobering thought or should be for today's modern civilization and strengthens credence of what an event could have done to an ancient civilization as proposed by Dr. Schoch.

Allow me to somewhat assuage any fears for present-day civilizations. The U.S. government has paid attention to the ramifications of this 2012 event. In 2016, President Obama signed an executive order to prepare for and protect against such a future event. Below is an abbreviated and abridged version of the order.

Preparing the Nation for Space Weather: New Executive Order
October 13, 2016 at 10:00 AM ET by Dr. Tamara Dickinson

Today, President Obama signed an Executive Order that seeks to coordinate efforts to prepare the Nation for space weather events. The Executive Order will help reduce economic loss, save lives, and enhance national security by ordering the creation of nationwide response and recovery plans and procedures that incorporate technologies that mitigate the effects of space-weather events. By this action, the Federal Government will lead by example and help motivate state and local governments, and other nations, to create communities that are more resilient to the hazards of space weather.

The term "space weather" refers to effects on the space environment that arise from emissions from the sun, including solar flares, solar energetic particles, and coronal mass ejections. Space weather is a natural hazard that can significantly affect critical infrastructure essential to the economy, social wellbeing, and national security, such as electrical power, water supply, health care, and transportation. *(Author's emphasis.)*

These emissions can interact with Earth, potentially degrading, disrupting, or damaging the technology that forms the Nation's backbone of critical infrastructure.[76]

It can be seen clearly that such a CME earth "hit" would be a devastating event, and the U.S. Government considers it more than just a theory (thank goodness). An event that can occur more often than previously realized, which then increases the possibility that such event occurred circa 9700 BC with devastating effect as Dr. Schoch theorizes.

The second theorized possibility of an extraterrestrial event (comet/meteor strike) that could have caused the disappearance of ancient cultures that may have thrived prior to the period of about 11000 BC to 9500 BC is called the Younger Dryas Event (YDE) or the Younger Dryas Boundary event (YDB). Scientists have puzzled over and debated the sudden extinction of the megafauna and flora in North America in the period around 11,000 BC to 9500 BC, known as the Younger Dryas period; a time when global warming trends suddenly reversed into a sharp temperature decline over most of the northern hemisphere—North America, Western Europe, and into Asia. North America and Western Europe are important impact areas where there is evidence of the premised seafaring megalithic builders. The early primary theories debated for the megafauna and human culture disappearances were over hunting and the temperature returning to a glacial-like period. Neither could be well supported. Then in 2007, a new theory was proposed. It has been called the Younger Dryas Event (YDE) hypothesis or Younger Dryas Boundary (YDB) hypothesis.

Dr. Richard B. Firestone, a nuclear chemist, and Dr. James P. Kennett, Professor Emeritus, Department of Earth Science and a marine geologist, along with their team proposed that an extraterrestrial (ET) impact—either a meteor, asteroid, or comet, more probably a comet—hitting the earth or exploding in the atmosphere was the cause of the extinction event and global cooling. Their central evidence is widespread discovery in the Younger Dryas boundary line of a "black mat" line evidencing massive wildfire, and high concentrations of microspherules created by high temperatures of non-volcanic origin and other sedimentary evidence. This is the same type of evidence Dr. Schoch cites for his CME impact theory, which he states would have caused similar burning and high temperature effects. Since Firestone and Kennett's proposal, their theory has been hotly debated and contested. In Firestone and Kennett's case, time has been supportive of their YDB hypothesis with discoveries and research providing more collaborating evidence.

There are three such recent reports that provide significant supporting evidence for the YDB extraterrestrial impact hypothesis, two that almost came out simultaneously. One is the recent discovery of a massive impact crater under the ice of the Hiawatha Glacier in northwestern Greenland. The preliminary research reports provide the possibility it occurred approximately 9800 BC, putting it in the window of the YDB hypothesis impact, circa 11000 BC – 9500 BC. The initial report and its associated geologic evidence have already moved other scientists to the validity of the YDB hypothesis; there is still much research to be done, but this discovery definitely moves the needle. One example of this is Stein Jacobsen, a Harvard University geochemist who studies craters and states that the geologic evidence (platinum peak) has made him a believer again of the YDB hypothesis. He states: "It's got to be the same thing."

Dr. James Kennett, one of the original authors of the YDB hypothesis, on learning of the Greenland impact evidence states: "I'd unequivocally predict that this crater is the same age as the Younger Dryas."[77]

A second report with similar research at White Pond, South Carolina, cites the work done on the Greenland impact crater and states that their findings in South Carolina "…at White Pond are consistent with an extraterrestrial impact event that triggered widespread biomass burning, as observed globally and reported elsewhere during the YD; however, the severity of environmental disruption at White Pond, its role in local megaherbivore extinction, and its impact on human life are yet to be determined. …In summary, the combination of proxy evidence within core sediments currently **supports the cause-and-effect linkage of an extraterrestrial impact with large-scale regional biomass burning, abrupt YD climate change, and megafauna declines leading to eventual**

extinction."[78] (Author's emphasis.)

The third report supporting the YDB impact hypothesis and published nearly simultaneously with the Greenland impact crater report is almost as dramatic as the discovery of the Greenland crater impact itself; it shows solid evidence that this extinction-causing extraterrestrial event was much wider spread than first determined—beyond the Northern Hemisphere to the southern tip of South America!

This scientific report finds that the evidence found at the Pilauco, Chile, site correlates with the evidence of mass extinctions across the Americas. In their own words:

"The Younger Dryas (YD) impact hypothesis posits that fragments of a large, disintegrating asteroid/comet struck North America, South America, Europe, and western Asia ~12,800 years ago. The sudden disappearance of megafaunal remains and dung fungi in the YDB layer at Pilauco correlates with megafaunal extinctions across the Americas. The Pilauco record appears consistent with YDB impact evidence found at sites on four continents."[79]

The report's summary states that this is the first and also extensive evidence found at high altitudes in the Southern Hemisphere providing evidence that a YDB impact event affected both Northern and Southern Hemispheres. In the summary, another statement is very important relevant to the premise of this book: "...Evidence has been found in the Pilauco **section that is similar to that found at >50 YDB sites on four continents."** (Author's emphasis.)

Please note in the last quote that the evidence of such an impact has been found at over 50 sites and four continents. In another section of the report, the type of impact is defined more specifically.

"There is a reasonable probability of one or more encounters within the last 13,000 years with debris swarms from the Taurid Complex or other large fragmented comets, and such an encounter would be hemispheric in

scope, lasting for only a few hours. The resulting debris field would be a mixture of dust and larger fragments, potentially equivalent to the impact of ~1000 to 10,000 destructive airbursts, such as occurred in Tunguska, Siberia, in 1908. **If such an event occurred at the YD onset, larger objects in the debris swarm could have created craters on land, struck the world's ice sheets, and/or impacted the world's oceans, creating severe biotic and climatic disturbances."**[80] (Author's emphasis.)

First, so much has been shared on this journey—all the evidence, all the scientific and archeological papers, and the theory presented here within a 20,000-year-old lost civilization that shared the construction knowledge of building megalithic-style sacred structures for sacred sites and did this on a global scale. They brought the designs and engineering for these sacred sites to function as places to bring Heaven and Earth together for the raising of consciousness of the participants, even incorporating acoustics/sound to assist in their consciousness journey. The key measurement of these site designs is evidenced as a length of 27.5 inches, a light/spirit cubit. Further and as important for our journey and the journey of humanity, the evidence is present to posit this technology came united and with a spiritual philosophy symbolized in the raised serpent with its purpose to raise the kundalini energy through the spine. The length of the spine then was used as a codified unit of measurement with its deeper symbolism in the design of sacred sites. This voyage has come a long way from wondering about a 27.5-inch unit of measurement, hasn't it?

I sit back in awe of these scientists' discoveries and conclusions, and I also feel a sense of loss. They are describing a global catastrophic event with severe effects in both hemispheres on four continents and in the world's oceans. That is the geologic answer to the question of what happened. Their evidence of a global-wide "black mat" sedimentary line of destruction and the disappearance of the megafauna is sobering in its implications. What does not seem to be addressed—or is addressed in a quiet, low-key way—is what happened to the people, the human populations, and the potential civilizations. It's as if the magnitude of this question is grasped at another level and like people visiting a cemetery one walks through it quietly and carefully, speaking in hushed tones. It also reminds me that every day alive is a gift, a good reminder.

This sense of loss is not just welling emotions; it is a solemn realization that the science on the possibility of such human societal losses shows it may well be a reality. In objective terms it is called a population bottleneck. A 2005 science paper by Dr. Jody Hey, a professor of genetics at Rutgers University, using DNA analysis and other evidence concludes:

The overall picture that emerges is one in **which the New World was very recently founded by a small number of individuals (effective size of about 70).**[81] (Author's emphasis.)

The New World, the Americas, was populated, or repopulated, by a group of 70 people. Imagine that, the Americas and just 70 people. This is the effective size; there would have been some others, but we are still describing an area of two continents. Effective population size is basically the number of individuals that can participate in producing the next generation. These would be the people capable of having children and would be less the group's overall population. Dr. Hey's analysis postulates that this "founding" population came from Asia around 12,000 to 14,000 years ago.[82] His conclusions align with the concept of surviving immigrants from a global catastrophe struggling just to survive and starting over.

Another science paper based on DNA analysis written in 2012 brings this extraterrestrial impact and the global devastation such an event would wreak on human society into even more dramatic focus. This paper, written by Dr. Heng Li, a professor from Harvard University, and Dr. Richard Durbin, a professor at the University of Cambridge, England, is based primarily on DNA analysis. Li and Durbin report a severe human population bottleneck, this time not in the Americas of the New World but rather in Europe and Asia. Note these are the four continents where the science research cited earlier in the chapter show evidence of a cataclysmic extraterrestrial impact.

We infer that European and Chinese populations had very similar population-size histories before 10–20 kyr ago. **Both populations experienced a severe bottleneck 10–60 kyr ago**, *whereas African populations experienced a milder bottleneck from which they recovered earlier.*[83] *(Author's emphasis.)*

Li and Durbin's range of the population bottleneck includes the time frame of the extraterrestrial impact posited by the other scientific reports. Their report continues that **the European and Chinese effective population dropped to 1,200.** (Remember Dr. Hey's effective population drop for the Americas was to 70.) They date this drop to between approximately 40,000 and 20,000 years ago, which is somewhat off from the extraterrestrial event, but it is close, and their earlier quote in the report of the severe population bottleneck is between 10,000 and 60,000 years ago. The early proposed cataclysmic event remains in the window both in time ranges and impact. These two reports of population bottlenecks, a technical way to write of near-extinction events, occurred on the same four continents where earlier reports believe the comet impact had a devastating effect. This would be a good time to repeat an abbreviated earlier quote, and remember, I stated how these researchers seem to reference the human impact in a low-key way. When you read "severe biotic and climatic disturbances," know that *biotic* means all livings things, including people. The population bottleneck studies provide the specifics of just how severe the event was for the physical population; imagine the impact on the psyche of the survivors. I wonder what surviving such an event would do to me or my descendants, how it would be remembered generations later. Plato was more prescient than we thought.

"There is a reasonable probability of one or more encounters within the last 13,000 years with debris swarms from the Taurid Complex or other large fragmented comets, and such an encounter would be hemispheric in scope... If such an event occurred at the YD onset, larger objects in the debris swarm could have created craters on land, struck the world's ice sheets, and/or impacted the world's oceans, **creating severe biotic and climatic disturbances**."[84] (Author's emphasis.)

Besides the severe biotic event, their conclusion in the above quote that "there is a reasonable probability of one or more encounters within the last 13,000 years with debris swarms from the Taurid Complex or other large fragmented comets," is highly significant in the next section of this chapter where we move from the scientific evidence of geology and geophysics to an ancient megalithic cultural site and its symbolism whose age and interpretations of its symbolism support an extraterrestrial event at the Younger Dryas Boundary in dramatic fashion.

There is mounting geologic evidence that an extraterrestrial Earth impact event occurred, whether a solar coronal mass ejection or something along the lines of a comet strike. In both cases this possible/probable event occurred in the Younger Dryas geologic period (circa 11000 BC–9500 BC) and had a devastating global effect on the flora, fauna, and human populations. Such an event on any advanced and evolving society would have had the result of "bombing them back to the Stone Age." The geological evidence of such an event is powerful, and Plato's narratives note that such events have occurred and describe the results on civilizations. The evidence put forth through these chapters of an advanced society with contact over several continents and one that became a "lost civilization" due to this event is persuasive and a distinct possibility. As for finding much evidence of this lost civilization, other than the tales passed forward by the survivors, the chances are negligible. I would suggest watching a fascinating History Channel documentary *Life After People*.[85] Basically this documentary shows what would happen if all the people of Earth were suddenly removed from the planet, with no one left to manage towns, cities, and infrastructure of civilization, a late 20th century civilization. What evidence would be left? The determination was that after 1,000 years there would be next to no evidence left of any civilizations; imagine after another 3,000 to 5,000 years. Ultimately they theorize after 10,000 years, there would only be the remains of the Great Pyramid, the Great Wall of China, and Mount Rushmore—stone edifices. Now remember the earlier quoted scientific papers that estimated an "effective population" from a "human bottleneck" event; 70 people for the Americas and 1,200 people for Europe and China. Further this projected effective population's primary focus under such circumstances would be basic survival. The documentary focuses just on the disappearance of people and what time and nature would do to what was left. Imagine if this was exacerbated by an extraterrestrial impact as theorized by the experts. Probably the only indications of such a global cataclysm are the evidence and survivor's tales that have been presented here.

This final segment of "what happened" synthesizes both the geologic and societal evidence from one site: Gobekli Tepe in Turkey. Gobekli Tepe is an absolutely incredible archeological site that has really turned archeology on its head. The discovery and the ongoing research of this site have had a textbook-changing impact on the age and beginnings of "modern human civilization"; it has put it in an "all bets are off" column.

We all owe a debt of gratitude to German archeologist Klaus Schmidt. Gobekli Tepe had been cursorily investigated and basically written off, but Dr. Schmidt decided to further investigate (thank goodness). The excavations started in 1995 led to the uncovering of probably one of the earliest sites not only into humanity's thought processes and abilities but their spirituality. Sadly Dr. Schmidt passed unexpectedly in 2014. His team and the archeological community continue to honor him and his work, and excavations are slowly moving forward.

What has been discovered is a site of multiple megalithic structures similar to Stonehenge, only the structures date back to at least 10000 BC! That is 7,000 years earlier than the positioning of Stonehenge's pillars. To date four enclosures at Gobekli Tepe being excavated are a major focus of the work; they are circular style structures that are generally ringed with limestone T-shaped pillars along with two T-shaped central pillars in the center. The peripheral pillars average about 11.5 feet in height, and the central pillars reach over 16 feet in height. These pillars weigh up to 30 tons, and between these four enclosures, 46 such pillars have been discovered to date. That is to date: non-invasive geophysical investigation indicates the presence of 21 such circles with over 200 pillars. The depths of some of these sites provide the possibility that this site construction activity may go back over an additional 2,000 years. Further what is presently being excavated and investigated is only 5 percent of the entire site. The surface of this discovery and the information it will provide has been barely scratched. This site has been called the "Cradle of the Gods," "The World's Oldest Temple," and a "Garden of Eden" area, and, as itself, is just in its own genesis of discoveries.

Gobekli Tepe and all that is being discovered is so exciting, there are not enough words for it. One of my greater frustrations has been not being able to visit this site; personally, so much can be experienced and grasped by being at such sites, and they are inspiring. Sadly, and I am sure it's the same challenge many independent researchers face, having the resources for such a journey is difficult. Combined with the unstable geopolitical area and almost continuous travel warnings, my own travel remains in the future, but my hope springs eternal.

In the meantime if you find this as exciting as I do, there are some books and articles slowly coming out about this site that run the gamut of perspectives and will provide different fascinating views. I would also recommend a wonderful 2012 documentary by National Geographic about Gobekli Tepe called "Cradle of the Gods." This is one to watch. I have multiple times; you will not be disappointed.

Fig. 50 Gobekli Tepe: Turkey

Now, bringing Gobekli Tepe back into a more narrow focus for the premise of this book, Gobekli Tepe is an ideal example of the fusion of humanity's much earlier use of technology and of serpent symbolism. This site also provides evidence of the possible catastrophic extraterrestrial event that scientists and archeologists have proposed.

Until the discovery of Gobekli Tepe, most archeologists believed that megalithic construction 12,000 years ago was impossible, that it was a time of small human societal hunter-gatherer groups that did not have the population, the organization, the resources, or the ability for such engineering. Gobekli Tepe has proven this incorrect on a grand scale; remember the evidence indicates there are 21 circles with over 200 stone pillars. This was not a one-time anomaly; what is exhibited is a continuous long-term process and not a process of constructing some type of protective habitations, but a practice of creating multiple megalithic stone circles with carvings and symbolism. The carvings and symbols are elaborate and prolific enough that some theorize this may be the dawn of proto-written language. The evidence is not only of possible proto-writing but also as early astronomers marking important stars and heavenly alignments and possibly recording celestial events. This points to powerful and shared abstract and metaphysical evolved consciousness that appears, even at this early time, to have combined physical engineering skills with a developed spiritual philosophy, 7,000 years before Egypt's first dynasty, further in the past from ancient Egypt than we are further in the future from ancient Egypt. That is something to pause and think about.

As of yet I have not been able to find any data of specific measurements of the different enclosures still being excavated and whether evidence of a 27.5-inch unit of measurement may have been used in the designs. There are many enclosures yet to be uncovered, and time will tell. A recent paper published in 2017 by Dr.

Martin Sweatman and Dr. Dimitrios Tsikritsis of the University of Edinburgh addresses evidence at Gobekli Tepe of the ancient comet discussed here. This paper also discusses the serpent symbolism found at Gobekli Tepe along with the recording and apparent significance of the star Deneb by the designers of Gobekli Tepe, which I find pertinent to the discourse of this book.

Much dialogue has already been shared on the YDB cataclysmic event, and one of the primary suspects of this event has been identified as a possible comet from the area of the Taurid complex or meteor stream. This paper comes to the same conclusion; the difference is that, rather than geological evidence, this appears to be a recording of the actual event in real time. The people at Gobekli Tepe apparently observed it happen! Below are excerpts of their paper's findings.

> ***We find the symbolism at GT provides strong support for the Younger-Dryas (YD) event as a cometary encounter, and hence for coherent catastrophism.*** *P.234* (Author's emphasis.)

> ***The proposal that Göbekli Tepe was, among other things, an observatory for monitoring the night sky, especially the Taurid meteor stream****, because of the disastrous consequences of the YD event appears to be the most complete and consistent interpretation of its symbolism yet developed. Certainly, no other interpretation has the level of statistical support described here.* P.245 (Author's emphasis.)

> We therefore conclude that the probability that Pillar 43 does not represent the date 10,950 BC is around one in 100 million, or one in 5 million if we neglect permutations with repeated symbols on Pillar 43. [86]

This is an amazing conclusion. This megalithic site has been compared to Stonehenge but was built 7,000 years earlier and thus may be the progenitor or early descendant of the global seafaring culture of such sites posited in this book and apparently observed the proposed YDB event impact. In the early cited paper of the YDB event evidence found in Chile, over 50 sites with impact evidence have been discovered so far. The article provides a map with the vast majority of the sites being in Western Europe and North America. Gobekli Tepe in Turkey would have been impacted also but not as severely or as immediately. I cannot even imagine what people would have seen in the sky or felt as it occurred; in any case I think it would have been dramatic and overwhelming, certainly worthy of documenting to say the least.

The conclusion of this research is important enough in itself, but I also found important and pertinent to this proposal the details Dr. Martin Sweatman and Dr. Dimitrios Tsikritsis provide in coming to their conclusions. One such detail is their observations in reference to serpent symbolism. An overall view of serpent symbolism has been shared along with its application to the previous sacred sites and its symbolism as it relates to the spine and the light/spirit cubit. Their report in reference to serpent symbolism at Gobekli Tepe was not unexpected, yet at the same time still surprised me; I think you'll see why.

"Snakes are the most prevalent motif at GT, although none appear on pillar 18 itself" (p. 242).

According to Peters & Schmidt (2004), **"While the snake/serpent is the most popular motif by far, no snake animal remains have been found.** This contrasts with all the other animal symbols for which copious animal remains have been found. In fact, the most prevalent animal motifs are, in order (snake, fox, boar, crane, aurochs). In terms of animal remains we have, in order (gazelle, aurochs, wild ass, mouflon, fox, boar). Even fish remains have been found, but not a single snake. **This singles out the snake symbol as potentially having a different meaning to the other animals.** Given the many threatening postures assumed by snake motifs at GT, the relationship snake/serpent = death and destruction is viable, but far from certain. Comets are certainly dangerous and destructive. Moreover, the serpent motif is a good symbolic representation of a meteor track." [87]

In the four sacred sites discussed in this book (Great Pyramid, Kukulkan Pyramid, great kivas, and Stonehenge), the importance of serpent symbolism was clearly shown in each. The probable use of a unit of measurement based on the spine and representing the spiritual serpent energy was also shown. At the same time none of these four sites had serpents outright depicted other than the Kukulkan Pyramid, named for the Maya plumed serpent deity with serpent heads at its base and designed to create a serpent of light on the pyramid during the solar equinoxes. At Gobekli Tepe serpents are "the most prevalent motif." Of all the multitude of animals depicted, the remains of all of them have been found on-site—except for serpents. This gives the serpent and its symbolism a highly unique status. Where Peters and Schmidt propose a connection here of death and destruction and/or a meteor track to the serpent symbolism, they also add the possibility a connection to the Greek constellation Ophiuchus (Serpent Bearer). I perceive a much different interpretation. Yes, the serpent can be dangerous; as I have explained, the serpent is a representation of positive spiritual consciousness energy, provided it is a raised consciousness energy. I believe the fact that the only remains not found are of serpents is evidence of the high esteem and sacredness serpents were afforded. As an example of this perspective, I provide an image of a limestone carved sculpture of a human head with a serpent on it. This was found at another archeological site near and related to Gobekli Tepe called Nevali Çori and dated to approximately 8500 BC.

What does this mean for ancient societies? What happened?   173

Fig. 51 Nevali Cori Serpent Head

I think it is clear in this sculpture, this is not a threatening serpent. I see a clear representation of the spiritual "serpent energy" rising up to the crown of the head and reaching higher consciousness. It is a depiction of bringing Heaven and Earth together in one's consciousness. If you look closely at the carved serpent of this Nevali Cori head, you notice the serpent head is triangular and two sides of the triangle are accented, highlighting the "V" of the triangle. Below is a statue from ancient Egypt with a similar serpent style on its head which aids in the suggested context.

174   The Spirit of Light Cubit

Fig. 52 Close up of Henu statue serpent. MMNY

Fig. 53 Egyptian Henu statue. MMNY

You can find this statue in the Metropolitan Museum of New York. Dated to 350 BC in Egypt, it is a Henu ritual figure and in "probably the final pose in a ritual dance of praise and jubilation called the Recitation of the Glorifications." The MMNY description also states, "...the figure carries a triangular-headed serpent over its forehead; this emblem is found on a restricted number of royal images."

I want to bring to the reader's attention that the serpent on the statue's head, rather than being the expected depiction of a serpent head on the vast majority of Egyptian busts here it is simply triangular/V-shaped. The MMNY notes such triangular head type serpents are limited to royal images. Earlier it was seen the images of the serpent rising from the forehead in so many Egyptian are visibly different. The triangular head serpent is linked to the Egyptian ritual of glorification.

> *Its goal is the transfiguration of the ka-spirit of a deceased person into a deified state, into being an akh, that is, a transfigured and equipped spirit of light. Egyptologist Mark Smith sums up...the aim of securing their elevation to a particular state of existence.* **Important features of this elevation are the complete restoration of mental and physical faculties and integration with hierarchy of gods and blessed spirits.**[88]

"The complete restoration of mental and physical faculties and integration with hierarchy of gods and blessed spirits"; the physical is restored and united with the Divine. This aptly describes uniting Heaven and Earth within oneself. I believe the Nevali Cori head has the same type of symbolism. This is serpent symbolism at its most positive and fits with the prevalence of the serpent symbolism and their apparent revered status at Gobekli Tepe and, as has been shown, around the world.

It also appears, with the available evidence, that the serpent symbolism was not just confined to the physical site of Gobekli Tepe. Dr. Martin Sweatman and Dr. Dimitrios Tsikritsis have also noted the star Deneb in the northern sky seems to have been a focal point for the people of Gobekli Tepe. The Deneb star has been noted by other researchers in similar manners. Independent researcher Andrew Collins was an early proponent of this and the significance of Deneb being part of the Cygnus constellation; Cygnus depicting a swan, which at the time would have been circumpolar stars, stars that did not set below the horizon. (See *The Cygnus Mystery*, 2006 and *Göbekli Tepe: Genesis of the Gods*, 2014.) I agree with their conclusions that the star Deneb was important at Gobekli Tepe and at other sites, as was the constellation Cygnus. Please note that the Cygnus constellation comes from the much later Greeks, but it is not uncommon for varied and earlier societies to imagine constellations similarly. We do not know how the people of Gobekli Tepe viewed it in any certainty.

From my research I propose that the view and stellar symbolism in their time included Deneb and Cygnus but also encompassed a larger stellar symbolism. In Chapter 3, I discussed the cave paintings of Lascaux and archeoastronomy expert Dr. Rappenglueck's interpretation of a 17,000-year-old cave painting he states depicts the star asterism; the Summer Triangle, which consists of the stars Altair, Vega, and Deneb—all circumpolar stars during this time. I believe this triangle is a key aspect in Gobekli Tepe heavenly symbolism and for many such sites. The immediate previous section showed the importance of the serpent with a triangular head. This Summer Triangle of stars, which includes Deneb, is the head of a stellar serpent of light that spans the sky, with its body being the Milky Way. (See the following image.)

What does this mean for ancient societies? What happened?   177

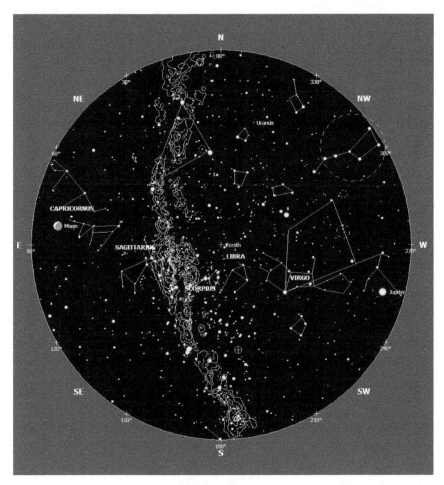

Fig. 54 CyberSky 5 planetarium program Turkey 12/21/11000 BC

The image shows what the star viewers at the time of Gobekli Tepe would have seen at night. I believe they would have interpreted it as a serpent of light with a triangular head spanning the sky from south to north with the triangular head created by the stars Deneb, Vega, and Altair being ever present in the northern sky. This is majestic and awe-inspiring; the image does not do it justice. Imagine looking up at the night sky, the heavens, and seeing this serpent of light encompassing and dominating the sky, not a serpent constellation made of some stars, but a serpent made of all the stars of the Milky Way. No wonder serpent symbolism is so prevalent through the ages and the world. (Note: CyberSky 5 is a software planetarium program available within most budgets. Since we cannot be there at 11,000 BC, the color program moving in "real time" and watching this stellar serpent move is a wonder and gives an appreciation of what early people were seeing.)

All this comes together even more elegantly. Remember, Deneb was in the constellation Cygnus the swan. Well, Vega is in the constellation Lyra, which depicts a vulture carrying a lyre, and vultures are also a significant motif at Gobekli Tepe. Next is the star Altair, which is in the constellation now known as Aquila, the eagle, which is likewise known as the flying vulture, again bringing it back in that aspect to Gobekli Tepe. The stellar serpent's head incorporates three bird constellations and in doing so creates a winged/plumed serpent! It is all there and fits seamlessly for Gobekli Tepe and all the symbolism and spiritual philosophies expounded upon in this book. This cannot be happenstance.

It actually, incredibly so, goes further. In using the planetarium program for different locations, Egypt, Mexico, England, and United States New Mexico, you can see that the people at these sites due to their more northern or southern locations will have somewhat different views of the night skies. In the cases of Egypt and Mexico being in farther southern latitudes than the others, while they will still see the Summer Triangle in the northern sky, they are able to see more of the southern stars than the other sites. This means that the Egypt and Mexico sky views included another asterism in the southern sky known as the Winter Triangle, consisting of the stars Sirius, Procyon, and Betelgeuse. In both Egypt and the Maya Mexico, Mesoamerican night skies, ancient stargazers would see not only the Summer Triangle of stars in the north but the Winter Triangle in the south, with both star triangles having the Milky Way going through them and on occasion both triangles in the sky at the same time. In either case these stargazers would see the Milky Way serpent body ending in a triangular head both in the north and the south. This being the case I think it is much more than coincidence that both the ancient Egyptian and Maya cultures have important serpent symbolism with heads at each end of the body (Nehebu-Kau and the double-headed serpent bar, respectfully) as an important aspect of their spiritually philosophies in connecting Heaven and Earth together (See Chapter 5). I imagine what all these ancient people saw in the heaven at night with no light pollution observing this serpent of light stretching from horizon to horizon; how could it not inspire any psyche.

Reluctantly I return from the magnificence of the ancient night sky back to the grounds of Gobekli Tepe, where there is one more important aspect to be explored. In Dr. Martin Sweatman and Dr. Dimitrios Tsikritsis' paper cited here, they also note an apparent proto-writing on the GT pillars, specifically what appear to be H's and V's on pillar 43, enclosure D.

"…we provide an interpretation for the abstract 'H-symbols' and nested 'V-signs' carved onto pillar 43. One possibility is that the H-symbols represent the position of Vega and/or Deneb, as both these stars would have appeared somewhat higher in the sky and slightly to the right (north) of the 'downward wriggling snake' (Serpens) around 10,950 BC. These bright stars would have been pole-stars in earlier millennia (Vega in *circa* 12,000 BC and Deneb in *circa* 16,000 BC)… (p. 239).

"… the 'H-symbols' demonstrate an early form of proto-writing existed at some point between 10,950 BC and 9,530 BC, at least for astronomical observations" (p. 243).

On the pillar they identify a "H"-like symbol as representing the stars "Vega and/or Deneb" next to a snake (serpent) and nested "V" signs on pillar 43 (the Vulture Pillar) and that it appears to be an early type of writing (alphabet) at least for astronomical observations with the V's possibly representing numbers. This brings us right back to the Summer Triangle of Deneb, Vega, and Altair and the proposal that they represent the head of a celestial serpent and all its metaphysical importance.

This conclusion is reinforced by a 2019 paper by Manu Seyfzadeh, Robert Schoch (yes, the same Dr. Schoch cited earlier) and moves us to pillar 18, titled "World's First Known Written Word at Göbekli Tepe on T-Shaped Pillar 18 Means God." I am sure this title got your attention as much as it did for me particularly that it focuses on the GT "H" symbol. In this case it was the "H" on pillar 18, also in enclosure D. In excerpts from their paper they state:

"We examined if H-shaped carvings in relief on some of these pillars might have a symbolic meaning rather than merely depicting an object of practical use. On Pillar 18 in Enclosure D, for example, one such "H" is bracketed by two semi-circles. An almost identical symbol appears as a logogram in the now extinct hieroglyphic language of the Bronze Age Luwians of Anatolia and **there it meant the word for 'god'"** (p. 31). (Author's emphasis.)

(Author note: The Luwians were a society that lived in the area of modern Turkey circa 2000 BC.)

"Pillar 18 rests on a pedestal with bird reliefs on its façade (Figure 3c). Besides a foxlike animal on its 'torso' (Figure 3b), it features a finely carved belt with several H-shaped symbols (Figure 3g & Figure 3h) and a buckle from which an animal hide loincloth hangs (Figure 3c). At the top front of the pillar is a set of three symbols composed (from top to bottom) of another **H-shaped symbol and an umbilicated disc hovering within the concavity of a crescent** (Figure 3d). Of note, the 'head' of the pillar is unmarked, though there are other pillars whose topmost parts are ornately carved with animal and geometric motifs."[89] (Author's emphasis.)

Following are images of pillar 18 being discussed.

What does this mean for ancient societies? What happened? 181

Fig. 55 "HU" close-up on Pillar 18

The image of pillar 18 shows the "H" previously discussed. In such "proto-writing" symbols, much more could be discussed. I just want to point out that between these two papers, the "H" relates to the stars Deneb and/or Vega and/or the U/V symbol also described as a crescent is linked to it. Such a link makes this a potential "HU" or "HV" symbol. We are stretching into possibly some of the earliest beginnings of modern alphabets; a nebulous affair, as a more recent example the differences between the letters "U" or a "V" being variations of each other and still evolving through the Middle Ages. Even that being the case, I take you back to Chapter 5 and Stonehenge.

More important and central to the Druid spiritual philosophy than the healing serpent eggs is the Druid deity HU, whose symbol was the serpent, as explained by Manly P. Hall:

Their temples wherein the sacred fire was preserved were generally situate on eminences and in dense groves of oak, and assumed various forms—circular, because a circle was the emblem of the universe; oval, in allusion to the mundane egg, from which issued, according to the traditions of many nations, the universe, or, according to others, our first parents; serpentine, because **a serpent was the symbol of Hu** (Author's emphasis) *the* **Druidic Osiris; cruciform, because a cross is an emblem of regeneration; or winged, to represent the motion of the Divine Spirit.**[90]

Please read Manly P. Hall's quote again. This is another epiphany moment or in today's text messaging an OMG. Here, from Stonehenge, the other example of megalithic stone circles and the Druids, comes the serpent as a symbol of HU—the proposed proto-writing at Gobekli Tepe representing both the celestial serpent and the divine. Further, the quote continues, noting the cross is a symbol of regeneration (rebirth), being the winged (plumed) motion of the divine spirit. I am running out of adjectives; this is stunning. Also the Cygnus constellation has within it another well-known asterism of stars called the Northern Cross, with Deneb being the pinnacle. Finally, as one author describes the Northern Cross stars, they are the "backbone of the Milky Way." The Northern Cross serves to point out the Milky Way—the luminescent river of stars passing through the Northern Cross and stretching all across the sky.[91] The "backbone," the spine; it all fits. This is not hammering a square peg into a round hole; it flows effortlessly.

I understand this has been a long answer to "what happened," but when answering such a question spanning tens of thousands of years of humanity along with the rise and fall of civilization and global-impacting events with its influence to the human psyche and spiritual consciousness, it is hard to do justice to such an encompassing question, and for the readers this is actually the "abbreviated" answer.

Now, it's all been presented, the 27.5-inch unit of measurement I have called the light/spirit cubit and have proposed is derived from the length of the human spine as a body proportion. This measurement that was codified from the spine also was symbolic of the serpent and its metaphysical connotations of being a representation of raising the kundalini, the spiritual consciousness in meditative fashion; a way of bringing Heaven and Earth together in oneself. Found in the archeological records of Egypt, this same unit of measurement has been determined by archeologists to be used by the Ancestral Puebloans (Casas Grandes: Paquime). Then there is its kin, the Maya zapal measurement of approximately 55 inches, a double light/spirit cubit, and its other relative from Stonehenge, the megalithic yard of 81.6 inches, equivalent to a triple light/spirit cubit. The archeologists and scholars tell us this is evidence of communications between these cultures.

"Cultures sharing similar measurement systems likely had some form of contact. Should such a measurement be located in architectural remains, and appear in halves or doubles, then the probability that this measure reflects a real historical unit of measure increases."[92]

In each case persuasive evidence has been offered for the light/spirit cubit use at each cultural sacred site; sites that are axis mundis. The evidence provided significant numerical results for each individual culture. Further evidence was provided of the importance of serpent symbolism for each culture and how it related to raising the consciousness of the participants. All this further cements the idea that they "likely had some form of contact."

How this contact occurred was also addressed, whether over land or by sea; from the Maya to the Ancestral Puebloans and by ancient megalithic building seafarers from Europe to Great Britain onward to the Americas. Also presented was the probability of all these cultures sharing spiritual philosophies that incorporated serpent symbolism in similar manners and purposes. Piece by piece an answer to this mystery was woven together. The conclusion points to a technological and spiritual globe spanning cultures over 15,000 years ago. It can be called Atlantis, it can be called a proto-Solutrean society, it can be called a lost civilization, or it can be called civilization X; whatever the name, the evidence is there that it existed.

There is also the evidence of what happened to this mysterious civilization. Convincing data from archeological and geological studies provide the information of a global catastrophic extraterrestrial event that severely impacted at least four continents and the oceans. This evidence is not only in the sediment and the strata but appears documented in the stone pillars at Gobekli Tepe.

Gobekli Tepe—an incredible site which is older than the extraterrestrial event itself and whose occupants seem to have been real-time observers of this civilization and an almost humanity-ending event; remember the population bottleneck reports.

Could Gobekli Tepe have been an outpost or a spiritual center of this lost civilization, a center left having to watch the devastation and destruction of their society with just sparse, scattered populations remaining, waiting to hear from any survivors? This may answer another mystery, one of Gobekli Tepe. The archeology work shows that the Gobekli megalithic stone circles were carefully and purposefully buried around 8000 BC. After waiting for generations and trying to keep their civilizations together, did they think the lands west were recovering from the disaster and made the decision to do a Hopi-like migration? Did they bury their sacred site to protect and preserve it from another heavenly disaster or as a time capsule for future generations? Perhaps one of these scenarios was the case; then as a last bastion of their civilization, they fanned out in a westerly direction in hope and trepidation.

Until further evidence is discovered, this must remain speculation, but it is not idle speculation. I turn back to the biblical narratives of the Garden of Eden and the Tower of Babel discussed earlier in this chapter. Are they skewed memories over the millennia of the fall of this lost civilization whose spiritual philosophy was to raise their spiritual consciousness? I postulated that the Garden of Eden account was the improper raising of serpent energy spiritual conscious falling deeper into physicality, and being driven out by a celestial being. This is also similar to Plato, who states what happened to the Atlanteans in their fall was losing their altruism to material possessions. Then in the Tower of Babel account, a global civilization sharing one language attempted to build a tower to Heaven. The tower narrative has a similar theme of a global society who loses their way and rather than raising their spiritual consciousness, they raise a material physical tower which is then struck down from the heavens, and they are scattered. In both accounts they travel west, the same direction proposed for the migration from Gobekli Tepe.

Is this how this lost global civilization was remembered, how serpent symbolism became more negative? Is this how a disastrous global event coming from the heavens was explained? How to make sense of it in the survivors' psyches that from that night sky spanning stellar serpent, death and destruction rained down. That this lost civilization also lost its spiritual way and was punished by the divine to, in Plato's words, "make them more careful and harmonious as a result of their chastisement." I think this is a distinct feasibility.

So these migrants from the proposed lost civilization, migrating westward and bringing with them their remaining knowledge to the enclaves of survivors, repopulating the lands from this population bottleneck event, knew basic survival came first and assisted the recovery with their remaining knowledge so that not all was lost.

The hard evidence and confirming investigation of a multicultural shared unit of measurement of 27.5 inches are compelling and significant. The scholars and archeologists who study measurement tell us of the importance of a culture's measurements in understanding their consciousness in their physical, cosmological, and spiritual views. These experts also tell us that if a specific measurement is found being used in the architecture of different cultures, whether the measurement unit is used in multiples or even halves, this is strong evidence that these different cultures had solid communication or possible shared ancestors. These criteria have been met in this discourse.

The research provided here also showed how exquisitely this unit of measurement functions as a seconds pendulum, one of the most important apparatus for the accurate measurement of time in the modern evolution of timekeeping. A seconds pendulum's length was important not only to measure time and in the forefront of society's modern consciousness, but it was also the original recommended length for a proposed international unit of measurement of space. Is this coincidence or a collective unconscious recalling an ancient memory?

Quoted earlier the French Academy described the meter, creating a measure that "transcends the interest of any single nation, thereby commanding global assent and hastening the day when the world's people would engage in peaceable commerce and the exchange of information without encumbrance."

I describe the ancient measurement postulated and described here in its elegance, representing the spine and unifying not only time and space but also Heaven and Earth; a measure that transcends any culture's physical sites and is transcendental for human consciousness!

This discourse went deeper into the significance of such a measurement. The origins of early measurement were derived from body proportions and provided the evidence for a compelling argument that this light/spirit cubit was codified from the length of the human spine—measurements used at sacred sites. The spine is known as the physical channel for the spinal cord and the central nervous system to the brain; it is also known as the spiritual energy pathway, the kundalini serpent energy, to raise consciousness to enlightenment.

I hope I have provided a cohesive picture of our journey. I find the proposals put forth here suitably and elegantly fit. I find these proposals resonate at many levels. They show that from a very early time, earlier than believed, the consciousness of humans went far beyond their physical surroundings into the metaphysical and they grasped the fallacy of the separation of the two. Humans built sacred sites to facilitate lifting the veil from a consciousness limited to the earth and unite it with heaven (universal consciousness). This unit of measurement did more than transcend borders of culture; it is transcendental for human consciousness. Humans synthesized their purpose, their goal, in an archetypal measurement that represented all of this—the light/spirit cubit—the only kind of measure that could ever come close to being a true measure of humankind.

...If I had ever been here before on another time around the wheel
I would probably know just how to deal
With all of you
And I feel
Like I've been here before
Feel
Like I've been here before
And you know it makes me wonder...
"Déjà Vu" by Crosby, Stills, Nash & Young 1970

7

What is the meaning, the message, for today?

The prior chapter could be considered the conclusion of the premises put forth in this book but I think the readers, as I did myself, have one more question. What does this journey of ancient peoples and evolving their civilization mean for us today? What was their aim? Are there any messages for us from our ancient ancestors that can help us in the present? I think there are, and the primary one is spiritual.

Perhaps this book's proposals are written in a too personal fashion in presenting the evidence, not being academically detached enough, yet when I have had, on occasion, an opportunity to talk to such scholars and academics, I can feel their passion and drive they share outside of their academic papers' detachment. For my writing and these discoveries that for some will be controversial, I just couldn't be detached. There should be given much credit and appreciation to researchers, archeologists, anthropologists, and religious scholars, these unsung heroes who have toiled, literally, in the trenches to save history and share their discoveries and research.

Humanity is more than science and physiology, as a whole and as individuals; science is important, technology is important, the evidence of a lost civilization is important. I feel what is more important is the consciousness of human beings and what motivates us beyond just survival, what inspires us. Thomas Keating pointed out so well that when God asked Adam and Eve, "Where are you?" the question was about their consciousness, not their physical state.[93] Are we just concentrating on the physical and ignoring spiritual, higher consciousness? Looking at all of humankind's stories and mythologies, what are they about? What tales stand the test of time? What draws us is the hero's journey, the journey of the saints. It is the stories of the perseverance and nobility of a good person, the redemption of a fallen person, that encourages us. Whether it is the biographies of famous people—be they scientists, military, statespersons, or success stories—it is not so much their accomplishments as it is their journey of overcoming challenges and remaining true and good that motivates us. It is the stories of those who sacrifice for the greater good that moves us to be more. It is the stories of success of the human spirit. This hero's journey is so ingrained in our psyche, it is one of the main themes of every book, movie, or television series we read or watch. It is our spirit and a consciousness beyond the physical that makes all the rest possible.

I can only present what the meaning and message mean to me personally; each reader will have to decide for themselves. I see it as human spirit and spirituality. Though experts have defined spiritual philosophies and put them in different categories, I use the term "spiritual philosophy" rather than "religion" in that the concept of a spiritual philosophy is a more encompassing term without dogma. One may consider oneself belonging to a particular religion and yet within that belief still have one's individual, personal spiritual philosophy remaining inclusive, yet individual. This perception harkens back to Chapter 1, where I shared this concept: to know yourself, to be yourself, yet one with the whole.

One of the proposals I presented was that this global, ancient lost society supported a unified spiritual philosophy with a core symbolism using a raised or plumed serpent representation. This serpent symbolism represented the raising of consciousness through meditative-like practices. It is a proposed central focus of this society and their creating sacred sites. Again going full circle back to Chapter 1, the practice of meditation in whatever name a specific spiritual philosophy uses to describe it—this act of raising consciousness, to unite Heaven and Earth in oneself and thus with everyone—is an integral practice in spiritual philosophies from ancient times. Whether it be in the Sufis of Islam, the Kabbalists of Judaism, Mystic Christianity, the Buddhists, the Hindus, the Taoists, the "Cloud of Unknowing" written in the Middle Ages, Buddha, St. Bernard, St. Teresa of Avila, St. John of the Cross, Rumi, on and on, it is known as a path to enlightenment, Oneness, apotheosis; to become friends of God.

Science, with its own core tenet to be able to observe, measure, and repeat, and thanks to its technology, has caught up with the effects of meditation or a version of it called mindfulness. The good news is that under the title of mindfulness, meditation is now more accepted in the mainstream of society. Mindfulness practice classes are being offered by the military, corporations, and fire/rescue departments, just as some examples, to help members cope with stress and PTSD. This acceptance has helped to open up avenues for more scientific studies of the effects and benefits from long-term practice of mindfulness/meditation. The results that are being reported are incredible.

Most are probably somewhat familiar with the concept of meditation, and the idea that the practice brings the meditator a sense of peace and calmness in a very basic way is true. Though I went into more of an overview in Chapter 1, here I want to add some details first of the physical results on the brain from meditation. What researchers are reporting in their studies on the physiological effects on the brain from meditation are dramatic. Using the latest scientific technology, they can measure and observe the brain in minute detail and in real time. The findings being reported are of actual neurologic changes to the brain. Discoveries include an increase in volume and density of the brain, apparent slowing of the brain's aging, increasing its preservation, and in comparison to non-meditators, their brains appeared seven years younger. The studies even noted meditation appeared to have greater effect on particular sections of the brain. Two areas of the brain were specifically noted for growth and density: Brodmann's 9 and 10, part of the prefrontal cortex. Though these areas are not well understood, they are considered involved in higher executive functions such as intention, planning, calculations, memory, empathy, and emotions. It has been theorized that the development of these areas of the brain contributed to the evolution of modern human beings and that meditation increases altruism and in doing so aids in the survival of a species.

What all this show is that meditation improves and evolves the brain physically and the mind spiritually.

I would say Teilhard de Chardin is a perfect example of one who grasped that last concept perfectly and expounded on it 60 years ago in two books: *The Phenomenon of Man* and *The Future of Man*. Chardin was a Jesuit priest and a scientist, specifically a geologist and paleontologist, and was involved in the work of the Peking man, an early ancestor of modern humans. He studied the physical evolution of human beings, but he also wrote of a human consciousness evolution. He proposed the human consciousness was evolving to a stage he called the "Noosphere," where individual consciousness would be connected in a collective consciousness and then would continue to evolve to an "Omega point"; a God consciousness or universal consciousness (being in the fullness of God). Chardin believed in both the physical and consciousness evolution.

Meditation/mindfulness practices provide the bridge between science and spiritual higher consciousness. Through meditation what is found by science and experienced in spirituality is that perfect connection between the physical and metaphysical, an ideal fusion of Heaven and Earth. This kind of fusion is what truly advances civilization. I believe this was the greatest aim of this lost civilization and the most important message they have left us.

In the present we are facing many of the same challenges of such an ancient society as Plato wrote they began losing their core value of humanity by overvaluing material wealth. Plato named this core value of humanity happiness, Aldous Huxley named it love, and science calls it altruism. Today we also face potential global catastrophic events—whether it be a comet, a CME, global warming, or other environmental event, the possibility is there. On January 23, 2020, in the Bulletin of Atomic Scientists, the Doomsday clock, which symbolizes, since 1947, the level of perils to humanity, was moved to 100 seconds to midnight, the closest it has been to midnight since its inception.

Are we going to heed the call and pay attention to the message from this lost civilization? This is not a message of doom and gloom; this is a message to be our higher selves, to listen to the better angels of our nature, our true natures. In any case I believe consciousness continues beyond physicality (I recommend reading the book *Mind Before Matter* for an expert discourse on this), but we are in physicality for a reason; to evolve our consciousness. All of our journeys are the hero's journeys, it is why we read about them or watch videos of them; they resonate within each of us. It is our true nature. If we do not listen to the message of this ancient civilization or remember our true nature individually or as humankind, as we have been advised, "Those who cannot remember the past are condemned to repeat it."[94]

I began this chapter answering one last question; I leave this chapter asking one last question and a request. Periodically ask yourself what was asked in the beginning of humanity's consciousness. Where are you? If you have a spiritual philosophy, you then probably believe our consciousness is more than our physicality; even if you do not, I am sure you have a standard of ethics, of altruism. In any case remember to ask yourself, "Where are you?" The question makes a good compass for the journey.

I hope this provides a measure of light for your journey/quest.

And I know if I'll only be true
To this glorious quest
That my heart will lay peaceful and calm
When I'm laid to my rest
And the world will be better for this
"The Impossible Dream (The Quest)"
by Mitch Leigh, with Joe Darion 1965

Epilogue

In full disclosure I want to share details of a key aspect of the beginning of this/my/our journey of discovery. I decided to wait for the epilogue to avoid the creation of any possible presuppositions or subconscious preconceptions. My concerns were that if I had shared in the beginning where I first read about an ancient Egyptian 27.5-inch unit of measurement used in Egyptian temples and pyramids, its source would have a biased effect on the reader. I wanted the reader to see all the connections and evidence laid out first to make a decision on the premises and the possibilities offered.

I first read about this Egyptian unit of measurement in an Edgar Cayce reading. Edgar Cayce was an American psychic, or as many such people are often called today, an "intuitive." He was born in Kentucky in 1877 and died in 1945. A "reading" was a requested psychic reading from Edgar Cayce; in 40 years he gave over 14,000 readings. If eyebrows have gone up on this revelation, I ask the reader to pause now and think back on this book: Were the evidence and premises strong enough and validating enough for the presented theories? Now, if the answer was no, the initial source won't make any difference; if the answer was yes, well, things just got a little more interesting.

Edgar Cayce was unique. I do not know of any other psychic, except possibly Nostradamus, who has such detailed records. I noted earlier that he gave over 14,000 readings; these were recorded and easily add up to well over 50,000 pages of material covering thousands of subjects. All these records are open to the public through the A.R.E. (The Association for Research and Enlightenment) with its headquarters located in Virginia Beach, Virginia. Approximately two-thirds of all the readings were medically related, and the records show there was great success in providing relief for a variety of conditions. Edgar Cayce is even considered the "Father of Holistic Medicine," incorporating mind, body, and spirit in the recommended treatment. Medical advice was the vast majority of requests for the first 20 years; it was in the next 20 years that readings on subjects such as ancient civilizations, past lives, our consciousness, psychic phenomena, and spiritual philosophies became more prevalent. For more in-depth and detailed information, I would recommend visiting www.edgarcayce.org.

The first book I read about Edgar Cayce was *The Sleeping Prophet*, a biography by Jess Stearn in 1978. As I related in Chapter 1, I was already into my journey of trying to understand science and spirituality. When I read about Edgar Cayce, I was thunderstruck—not about him being a psychic, but from the information provided from the readings. In my opinion it was the most complete and cognizant spiritual philosophy I had read; it put so many pieces of the puzzle together. I had never heard of this person, yet compared to the college courses I took on comparative religions and philosophy, along with such books I was reading on my own, the spiritual philosophy that came out of these readings rang the truest. I scoured bookstores looking for more Cayce material; I can tell you the availability was sparse. For those who remember the bookstores of the '70s, they had no category for the type of material found in the Edgar Cayce readings. You had to go to the "Occult" section (really). Fortunately, times have changed, though I mourn the loss of so many bookstores and being able to wander through the aisles looking for unexpected finds.

I did find more books on Edgar Cayce, and one was another biography, written by Thomas Sugrue, *There Is a River: The Story of Edgar Cayce*. Thomas Sugrue was a college friend of Edgar Cayce's son, Hugh Lynne, who, understandably, did not believe what Hugh Lynne shared about his father. Thomas visited the Cayce home during the summer and became convinced of the abilities of Edgar, like so many other initial skeptics. Later Sugrue became ill, and the readings from Edgar aided him in his illness. This all led to the biography on Cayce he wrote, and I recommend it, if for nothing else for the "Philosophy" section. You will not be disappointed. If this interests you I would recommend one other biography for its comprehensive detail: *Edgar Cayce: An American Prophet* by Sidney Kirkpatrick.

The few books I was finding at the time about Edgar Cayce material were really stunning; it was the Cayce books on spiritual philosophy and ancient civilizations that drew my attention. You could get a membership to the organization and receive a newsletter, research articles, and a list of the latest books about research of the reading materials. Needless to say I joined; this was the '70s Internet equivalent. I was so amazed that around 1979, I went to Virginia Beach to see if this place was "the real deal"; it was. They not only had an incredible library, and still do, on all these subjects; they had a bookstore with a wealth of research and writing on subjects from the Cayce readings. I left some clothes behind to make room for books in my luggage. After going home I continued to keep up to date from afar until I retired from my career and moved there. Yes, I did, for eight years, with the knowledgeable people, the library, and the conferences with both local and international experts on these subjects and more. It seemed inevitable for my journey.

This is enough Edgar Cayce background for now; their website can provide much more information. By 2005, I had read many of the books and articles on the Cayce material and had the opportunity to go to conferences and participate with local groups who also resonated with the readings' spiritual philosophy. At the same time I was drawn to how scientific discoveries in many different arenas were validating what was in the readings. I found it all fascinating, particularly the validated discoveries, which, for me, gave more weight to the spiritual philosophy that came from the readings. Almost everyone has a belief in some version of a spiritual philosophy, and these beliefs are generally reinforced by one's faith. I understand and respect that, yet, for myself, I am a bit of a doubting Thomas; that is my personal challenge. Seeing a wide variety of beliefs each with their followers and all confident in their views that their spiritual philosophy was the most correct, I needed more. Even as confident as I felt about the readings' philosophy and how much it had strengthened my spiritual beliefs, I knew it was still a spiritual philosophy among all the rest. I felt the validation of other Cayce readings gave increased support to the spiritual philosophy described in the readings. If you go on their website you will see seven such validations of the readings on history and discoveries; they are called predictions that came true.

Such validation gave me what I call a knowledgeable faith in spiritual philosophy and strengthened my convictions. I still pored over the readings to give myself better understanding and depth; to understand. I also was interested in more information on what the Cayce readings said about ancient civilizations, and this led me to Egypt. I think it was about 2005 when I stumbled across a Cayce reading on ancient Egypt that, almost as an aside, mentioned a unit of measurement used for the design and construction of their temples and pyramids. I have put excerpts of the reading below. The reading occurred in July 1935.

"You will describe in detail this temple ... It was in the form of the pyramid, within which was the globe—which represented to those who served there a service to the world.

"Here it may be well that there be given a concept of what is meant by the journey, or what journey is meant. As indicated, it, the globe within the pyramid without, was four forty and four cubits *(twenty-seven and one-half inches was a cubit then, or a mir [?] then)*. (Author's emphasis).

"…Hence man with his free will makes for whether the body is AS the temple of the living God or the Temple…

"As ye have seen in that thou hast given, may give, that led the many, that opened the way to many, it, the temple, thy temple, thy body, thy mind, thy portion of the God-Consciousness may be aroused and awakened to the abilities within self to assist in those becoming aware of the necessity of arousing to the destinies of the body, the mind, the soul." Reading 281-25 (Author's note: The Cayce readings are numbered to protect the confidentiality of the person asking the questions or to categorize questions in a specific area.)

In the larger body of the reading, that simple statement caught my attention, a statement mentioning, like an afterthought, an "oh by the way," the unit of measurement used was 27.5 inches. It was so specific I thought it would be easy to check on and research. Well, now the reader knows where that went. Incredible isn't it and such validation of information in the Cayce readings on so many levels.

As I researched this measurement through the years, continuingly finding more and more amazing connections and validations of it origination, use, and purpose, the larger body of the reading became more significant. In the excerpts shared here, the Cayce reading states that those at this pyramid-shaped temple were there for service to the world. The reading compares this physical structure temple to the human body as a temple and that its purpose was to raise participants' spiritual consciousness. As I have written, the Cayce readings are incredible; talk about finding validations! Documented in this Cayce reading are some of the same findings that this book and its research have reached. Only I took years and hundreds of pages to do so. In each example science and spirituality are being united. With what I have written here, I still ponder this reading.

I hope the reader can see after reading this book just how stunning and significant this reading has come to be, and it is not the only reading that I have found being validated. I have discovered a total of six, so far. I will share the two that pertain to archeology and the Maya culture.

In May of 1943, Edgar Cayce gave a life reading for a woman seeking to better understand her role in the universe. In his meditative-trance state, Edgar Cayce reviewed some of her past lives that would be of assistance to her in her present life's journey. This was always a core tenet of the life readings for such seekers. These readings were not for entertainment; they were for growth of the individual. This is clearly stated in the reading:

"In giving the interpretations of the records as we find them for this entity, these we choose from same with the desire and purpose that this information may be a helpful experience; enabling the entity to better fulfill those purposes for which the entity entered this present sojourn." (Reading 3004-1)

Her past life in Yucatan, described in her Cayce reading, mentioned an ancient Maya city that he called **Ichakabal.**

"Before that the entity was in the Yucatan land, when there were those activities in which there were those groups that had caused dissension among the worshipers in the temple there of **Ichakabal.** In those activities we find whole groups of individuals being separated, and seeking for activities in other groups.

"The entity maintained that the activities there, in **Ichakabal,** were to be kept." (Text of Reading 3004-1) (Author's emphasis.)

Now, more than 50 years later, the Maya city named by Edgar Cayce in his meditative-trance state is discovered! This is not a minor discovery; the researchers describe this site as a very important Maya site with a pyramid temple that absolutely dwarfs the Kukulkan Pyramid at Chichen Itza.

"The archaeological site in question, located 40 kilometers west of the city of Bacalar, was discovered in 1995 and everything indicates that this is a political and religious center of great importance in the Maya world, with a backbone of more than 40 meters high and 200 meters base, surpassing even the famous Chichen Itza pyramid."[95]

One of the aspects that really caught my attention was how this Mayan site discovered by the archeologists was given its name.

"The site is located west of the Bacalar Lagoon, surrounded by a lot of smaller and very close to the archaeological site of prehistoric settlements Dzibanché. **The name was assigned Ichkabal only in March 1995, when the first archaeologists came to the place accompanied by local guides**; the name means 'between low,'...intended to highlight the physiographic features of the site environment."[96] (Author's emphasis.)

The archeological site was named Ichkabal over 50 years after the Cayce reading named and identified it!

These quoted articles are from Mexico's National institute of Archeology and History. It should be noted that the readings spell this Maya city *Ichakabal*, while the archeologists spell it *Ichkabal*. I would suggest, since the readings were written down as Edgar Cayce spoke, that the slight difference is merely a phonetic variance.

The Cayce readings not only cited a past life in the Maya city of Ichkabal, but named and identified it over 50 years before its future discovery. More to ponder.

The next example is from a Cayce reading that stated there were Maya sculpted stones that were magnetized. This has also been validated. The Cayce reading is as follows:

"1. HLC: You will give an historical treatise on the origin and development of the Mayan civilization, answering questions.

"2. EC: Yes. In giving a record of the civilization in this particular portion of the world, it should be remembered that more than one has been and will be found as research progresses.

"13. Hence each would ask, what specific thing is there that we may designate as **being a portion of the varied civilizations that formed the earlier civilization of this particular land?**

"14. **The stones that are circular, that were of the magnetized influence upon which the Spirit of the One spoke to those peoples as they gathered in their service...**" (Author's emphasis.) TEXT OF READING 5750-1 circa 1933.

There has been renewed interest in the stone carvings with magnetic properties originally discovered by Dartmouth researchers in the 1970s (V. Malmstrom) and further evaluated in 1997. Recent studies (Harvard 2019) have confirmed that statues in the Guatemala area were purposely carved to focus their magnetic aspects in specific areas, usually the navel, forehead, or right cheek. There is also a large "turtle head" table-like sculpture with a northern polarity aligned at its snout and a southern polarity aligned at the back of its head. The archeologists debate the meanings for this but agree the sculptors recognized the magnetism found in stone (possibly caused by lightning strikes) and also naturally occurring magnetism in stone. The key point, relating to the Cayce readings, is that these sculptors and their culture had discovered a way to recognize and identify this magnetism and shape their sculptures accordingly.

The Edgar Cayce readings predate these university discoveries by over 40 years. The Cayce reading on the magnetized stones contains even greater accuracy confirmed by the academic researchers. The recent archeological findings further state that these magnetized stone carvings date to Olmec and Pre-Olmec societies, earlier societies that are considered ancestral to the Maya society. The Cayce reading was focusing on the Maya culture yet stated: "Hence each would ask, what specific thing is there that **we may designate as being a portion of the varied civilizations that formed the earlier civilization of this particular land**?" (Author's emphasis.)

Additionally, the Cayce readings relate that the magnetism designed in the stones is related to their spiritual belief; this possibility, though debated, is also put forth by the academic research.

"If the sculpture depicts a head, it is often magnetic in the right temple. If it depicts a body, its magnetic pole is usually near the navel...

"Clearly, something in the early Soconuscan culture seems to have dictated a linkage between the right temple and magnetism and between the navel and magnetism. What was it?

"Naturally, one might conjecture that the connection being implied between magnetism and the head was the symbolization of a mental or spiritual link... Similarly, the association between magnetism and the navel may well have been a commemoration of the physical side of life—the continuity of the life-force from mother to child, despite the cutting of the umbilical cord at birth."[97]

This ability to identify and recognize magnetic properties in stones and shape them for their own purposes in Mesoamerica and the Guatemala area was unknown to academia until the 1970s, yet the Cayce readings over 40 years earlier clearly cited this ability of the ancient Mesoamericans and their reasons for doing so. We are provided with incredible and further hard data evidence and validation of the accuracy of the Cayce readings in many arenas. I provide links below for your own interest.

https://www.livescience.com/65410-magnetized-potbelly-sculptures-guatemala.html

https://www.ancient-origins.net/news-history-archaeology/magnetic-stone-figures-0011870
https://www.dartmouth.edu/~izapa/CS-MM-Chap.%203.htm

Needless to say this doubting Thomas has been humbled. There is more here in the Cayce readings than meets the eye. The material is worthy of exploring at the very least. Science, archeology, physics, cosmology, neurobiology, etc. seem to be catching up to the information in the Cayce readings. As Shakespeare stated so well in Hamlet, "There are more things in Heaven and Earth, Horatio, than are dreamt of in your philosophy." Yes, our knowledge is more limited than we realize.

The spiritual philosophy is so unifying and cohesive; it provides answers on so many planes and at the same time allows a seeker to pursue their journey their way. It almost insists on it. The prescience of the readings should have been expected; they warned that the information was to assist the seeker in their own journey in their own spiritual philosophy and that the Cayce organization should never become or be considered any kind of sect; the material was to assist, not replace. I think of one such reading that gets this point across well. When asked about all the different version of the Bible available which version was the most correct, the answer was "The one you read." It was not the difference of spiritual philosophies or creating new ones; it was realizing the relationships, the unity between them, as exampled in this reading.

- Not their differences, but their overlapping. And what these teach in the many varied sects, the many varied cisms—not their differences but their unity. These will not only enable the entity to give to others but in the giving, in the seeking, in the understanding, broaden—yea, magnify—the vision and the ability in spiritual things. TEXT OF READING 1473-1

For me it is ultimately the spiritual things, the spiritual journey in a physical world. The readings stressed a wonderful practice to aid in this journey: meditation. I had noted throughout this discourse the importance and theme of this raising of consciousness. I also shared that I have been a daily meditator for 17 years now and it has made a wonderful difference in my life, as the neuroscientists are confirming now. I started because of its noted importance and recommendation in the Cayce readings and I couldn't put it any better.

- There are others that care not whether there be such things as meditation, but depend upon someone else to do their thinking, or are satisfied to allow circumstance to take its course—and hope that sometime, somewhere, conditions and circumstances will adjust themselves to such a way that the best that may be will be their lot.
- It is not musing, not daydreaming; but as ye find your bodies made up of the physical, mental and spiritual, it is the attuning of the mental body and the physical body to its spiritual source.
- TEXT OF READING 281-41

Of course me, being me, I had to get six different books about meditation and how to meditate, and after all that it was the simple method offered in the Cayce readings that worked the best for me. The daily meditation is a straightforward practice toward reaching our higher selves and spiritual values. Remember to complete the circle with the application of our higher self, thus avoiding a fall.

My journey to date with the Cayce readings and the A.R.E. includes my writing articles for their magazine, the A.R.E., and 4th Dimension Press publishing my first book. I have also lectured at conferences and led tours to some of the sites documented here. I am also a former member of their Board of Trustees.

I wasn't kidding about full disclosure. Ultimately it is you, the reader, who must judge the strength and validity of the premises presented of a shared ancient unit of measure, a global lost civilization, and its constructions and spiritual purpose to raise human consciousness, as well as the Cayce readings. I will share a personal experience from my journey and following this path that provided me the greatest confirmation that my "compass" is reading true.

The highest and humblest point and best example I can give of what the Cayce materials provided by the A.R.E. have done for me in my ongoing journey are the honor and words granted to me by my youngest son as he prepared to marry. He asked me to officiate their wedding. I was stunned and stammered, "Are you sure about this? Is everybody okay about this?" I had been divorced nine years earlier, so along with my son's fiancée's family, there were many logistics and I wanted to make sure their wedding was for them. He answered, "Yes, and, Dad, we couldn't think of anyone better. You're the most spiritual guy I know." I will never be able to describe my upwelling of emotions, and amid tears of love, I said yes to such wonderful, humbling words that I continue to strive to deserve and live up to. This personal story is not a testimony to my journey or me; it is a testimony to the spiritual philosophy in the Cayce readings and their application. I will always be grateful.

All written, it is up to each reader to decide whether the core premises of this book and the evidence meet the plausible possibilities threshold by themselves and/or the meditative higher consciousness evidence of the Cayce readings do also, whether individual or together. In any case I wish you all a journey of insight.

And you can fly
High as a kite if you want to
Faster than light if you want to
Speeding through the universe
"Thinking Is the Best Way to Travel" The Moody Blues 1968

"Your vision will become clear only when you can look into your own heart. Who looks outside, dreams; who looks inside, awakes."[98]

Post Script

*E*ven as this book comes to its conclusion, I continue to research ancient sites and dusty archives of archeological metrology for evidence of this shared spirit of light cubit, and it continues, appropriately, to come to light. I recently found a 1984 German research project report from the Central Archeological library of the Government of India department, New Delhi, India. It is a filed report on the archeological site Mohenjo-Daro. Mohenjo-Daro is an Indus Valley civilization, also known as the Harappan Civilization site circa 2500 BC, located in Pakistan, between India and Iran.[99] In this report is an analysis of the Harappan linear unit by R. C. A. Rottlander, starting on page 201, whose effort is to determine this culture's units of measurement. Rottlander determines the Harappans used an "Indus foot" of 34.55 cm and a "double foot" of 69.11 cm. These results are equivalent to the length of the spirit light cubit and its half. Asia can now be added to Africa, Europe, and North America as the fourth continent on which this unit of measure has been discovered. And so it goes. I am confident more such discoveries will occur.

Book Figures

1. By Provenzano15 - Own work, CC BY-SA 3.0, https://commons.wikimedia.org/w/index.php?curid=26261459
2. By Constantino Brumidi - Own work, CC BY-SA 3.0, https://commons.wikimedia.org/w/index.php?curid=134782)
3. http://en.wikipedia.org/wiki/File:Vitruvian_Man_Measurements.png#file.
4. https://commons.wikimedia.org/wiki/File:Paintings_from_the_Chauvet_cave_(museum_replica).jpg
5. https://commons.wikimedia.org/wiki/File:Höhlenmalerei-Lasc.png
6. https://commons.wikimedia.org/wiki/File:Simple_gravity_pendulum.svg
7. Author figure
8. https://commons.wikimedia.org/wiki/File:Akhet.jpg Creative Commons Attribution-Share Alike 3.0 license. Author: Luna92
9. https://commons.wikimedia.org/wiki/File:Leonardo_da_Vinci-_Vitruvian_Man.JPG Public Domain
10. Author's personal photo
11. Author's personal photo
12. Author's personal photo

13. By Yuchitown - Own work, CC BY-SA 4.0, https://commons.wikimedia.org/w/index.php?curid=45030146
14. Author's personal photo
15. Author's personal photo
16. Author's personal photo
17. Author's personal photo
18. Author's personal photo
19. Artist: Carol Hicks, commissioned by D. Carroll
20. Public Domain http://www.blackmesatrust.org/?page_id=46
21. https://commons.wikimedia.org/wiki/File:Fontanelle-dt.JPG
22. By Forest & Kim Starr, CC BY 3.0 https://commons.wikimedia.org/w/index.php?curid=71981387
23. This work has been released into the public domain by its author, Pearson Scott Foresman. https://commons.wikimedia.org/wiki/File:Spine_(PSF).png
24. Author's personal photo
25. Author's personal photo
26. Author's personal photo
27. http://commons.wikimedia.org/wiki/File:Y-Z.jpg   Author Sitehut
28. Spine—http://www.niams.nih.gov/health_info/scoliosis/.
29. Serpent—Courtesy of the author.
30. DoktorMax (talk) 16:31, 20 January 2008 (UTC) released to public domain
31. Author's personal photo
32. Source: https://www.flickr.com/photos/geraldford/7175237665/in/photostream/
33. Author's personal photo
34. Author's personal photo
35. Author's personal photo
36. By ATSZ56 - Own work, Public Domain, https://commons.wikimedia.org/w/index.php?curid=6344965

Book Figures   211

37. By Judson McCranie, CC BY-SA 3.0, https://commons.wikimedia.org/w/index.php?curid=57804011
38. Author's personal photo
39. Author's personal photo
40. Author's personal photo
41. Title: Annual Report of the Bureau of American Ethnology to the Secretary of the Smithsonian Institution Year: 1895 https://www.flickr.com/photos/internetarchivebookimages/14784501115/ Source book page: https://archive.org/stream/annualreportofbu219smit/annualreportofbu219smit#page/n477/mode/1up, No restrictions, https://commons.wikimedia.org/w/index.php?curid=43454946
42. This work has been released into the public domain by its author, Mr. Silva at English Wikipedia. This applies worldwide. https://commons.wikimedia.org/wiki/File:Valley_of_Fire_Petroglyphs_2.JPG
43. Author's personal photo
44. Author's personal photo
45. Author's personal photo
46. Author's personal photo
47. Public Domain: William Stukeley - http://www.avebury-web.co.uk/aubrey_stukeley.html
48. By William Blake http://www.blakearchive.org/exist/blake/archive/copy.xq?copyid=jerusalem.e&java=no Public Domain, https://commons.wikimedia.org/w/index.php?curid=26495185 https://commons.wikimedia.org/wiki/File:Jerusalem_The_Emanation_of_the_Giant_Albion_e_p100_300.jpg
49. By Darkone - Own work, CC BY-SA 2.0, https://commons.wikimedia.org/w/index.php?curid=28452
50. Author: Teomancimit https://commons.wikimedia.org/wiki/File:Göbekli_Tepe,_Urfa.jpg

51. Yonca Eldener Dr./Ph.D@YoncaEldener. 11:06 p. m. · 20 sept. 2018 Twitter for iPhone
52. Author's personal photo
53. Author's personal photo
54. Author's CyberSky 5 planetarium program set at Turkey 12/21/11000 BC
55. Author's depiction

   A: Book Cover: Author's creation using Fig. 9 and Fig. 28
   B. Back Cover: Pyramids, author's personal photo
   C. Bio image: Author's personal photo

Recommended Reading

Mind Before Matter: Vision of a New Science of Consciousness by Trish Pfeiffer (Editor), John E. Mack (Editor), Paul Devereux (Editor)

The Divine Within by Aldous Huxley

Aldous Huxley's Perennial Philosophy

Contemporary Cayce by Henry Reed and Kevin Todeschi

The Sleeping Prophet by Jess Stearn

There Is a River by Thomas Sugrue

Edgar Cayce: An American Prophet by Sidney D. Kirkpatrick

The Edgar Cayce Companion compiled by B. Ernest Frejer

Intimacy with God by Thomas Keating

The Complete Pyramids by Mark Lehner

Illusions by Richard Bach

Forgotten Civilization by Robert M. Schoch

Pyramid Quest by Robert M. Schoch PhD and Robert Aquinas McNally

The Future of Man by Teilhard de Chardin

Book of the Hopi by Frank Waters

Experiments in A Search for God: lessons 1-24 by Mark Thurston PhD

To Have or To Be by Erich Fromm

The Secret Teachings of All Ages by Manly P. Hall

The Cygnus Mystery by Andrew Collins

Forest of Kings by Linda Schele and David Freidel

The Human Condition by Thomas Keating

Chaco Astronomy by Anna Sofaer

The Archeology of Measurement edited by Iain Morley and Colin Renfrew

Author Bio

Donald B. Carroll spent his career working in Fire & Rescue, as a firefighter/paramedic, company officer, district chief, and an academy instructor. During those 30+ years, he raised a family and pursued the meaning of life through extended study into scientific, spiritual, and philosophical materials. He has been a regular speaker and writer of metaphysical topics, ancient culture symbolism, and a tour leader to many of the ancient sites he has written about. Much of his research includes spiritual symbolism and how it is shared across cultures and incorporated in sacred sites for the same perennial purpose.

Endnotes

1. A paraphrase of "That it, the entity, may know itself to be itself and part of the Whole; not the Whole but one with the whole; and thus retaining its individuality, knowing itself to be itself." (Edgar Cayce Reading 826-11).
2. "Turn on" meant go within to activate your neural and genetic equipment. Become sensitive to the many and various levels of consciousness and the specific triggers engaging them. Drugs were one way to accomplish this end. "Tune in" meant interact harmoniously with the world around you—externalize, materialize, express your new internal perspectives. "Drop out" suggested an active, selective, graceful process of detachment from involuntary or unconscious commitments. "Drop out" meant self-reliance, a discovery of one's singularity, a commitment to mobility, choice, and change. Unhappily, my explanations of this sequence of personal development are often misinterpreted to mean "Get stoned and abandon all constructive activity." Timothy Leary. *Flashbacks: A Personal and Cultural History of an Era*. Tarcher (March 17, 1997) p.253.
3. Albert Einstein. *Science, Philosophy and Religion.* (Sep 1940).
4. Aldous Huxley. *The Divine Within*. Harper Perennial Modern Classics (July 2, 2013) p.54.
5. Aldous Huxley. *The Divine Within*. Harper Perennial Modern Classics (July 2, 2013) p.71.
6. Kevin J. Todeschi and Henry Reed. *Contemporary Cayce*. A.R.E. Press (October 10, 2014).
7. Aldous Huxley. *The Divine Within*. Harper Perennial Modern Classics (July 2, 2013) p.15.
8. Flinders Petrie. *Measures and Weights*. London: Methuen & Company Ltd. 1934, p.1.
9. *The Archeology of Measurement: Comprehending Heaven, Earth and Time in Ancient Societies*. Cambridge University Press. 2010. Edited by Iain Morley and Colin Renfrew. p.1.
10. Marcus Vitruvius Pollio. *de Architectura*, Book III, Ch.1.
11. Frank Waters. *Book of the Hopi*. Penguin Books (June 30, 1977) p.9-10.
12. Frank Waters. *Book of the Hopi*. Penguin Books (June 30, 1977) p.4.
13. Flinders Petrie. *The Journal of the Anthropological Institute of Great Britain and Ireland*, Volume VIII. London 1879: p.107.
14. Flinders Petrie. *The Journal of the Anthropological Institute of Great Britain and Ireland*, Volume VIII. London 1879: p.107.
15. Flinders Petrie. *The Journal of the Anthropological Institute of Great Britain and Ireland*, Volume VIII. London 1879.
16. *Measure: Towards the construction of our world. The Archeology of Measurement: Comprehending Heaven, Earth and Time in Ancient Societies*. Cambridge University Press 2010. Edited by Iain Morley and Colin Renfrew. p.3.
17. G.J. Whitrow. *Time in History*. Oxford New York; Oxford University Press 1989. p.14.

Endnotes   217

18 http://news.bbc.co.uk/2/hi/science/nature/871930.stm Wednesday, 9 August 2000, 01:00 GMT 02:00 UK By BBC News Online science editor Dr. David Whitehouse.
19 Ken Adler. *The Measure of all Things*. Free Press; Reprint edition (October 1, 2003) p.89.
20 Ken Adler. *The Measure of all Things*. Free Press; Reprint edition (October 1, 2003) p.85.
21 Mark Lehner. *The Complete Pyramids*. Thames and Hudson, April 28th, 2008. p.130.
22 Sir Flinders Petrie. *Wisdom of the Egyptians* (1940) by Sir Flinders Petrie: 2019 digital edition. p.30-31.
23 Robert Schoch. *Pyramid Quest: Secrets of the Great Pyramid and the Dawn of Civilization*. Tarcher; 1st edition (June 2, 2005) p.139.
24 https://www.metrum.org/measures/index.htm 12/1/19 DBC A History of Measures Livio C. Stecchini First Published March 1, 1961 American Behavioral Scientist.
25 https://www.metrum.org/measures/index.htm 12/1/19 DBC A History of Measures Livio C. Stecchini First Published March 1, 1961 American Behavioral Scientist.
26 Mark Lehner. *The Complete Pyramids*. Thames and Hudson. April 28th, 2008. p.168.
27 Mark Lehner. *The Complete Pyramids*. Thames and Hudson. April 28th, 2008. p.168.
28 http://www.legon.demon.co.uk/metrorev.htm Article review by John A.R. Legon.
29 Flinders Petrie. *Measures and Weights*. London: Methuen & Company Ltd. 1934. p.7.
30 Flinders Petrie. *Measures and Weights*. London: Methuen & Company Ltd. 1934. p.8
31 Mark Lehner. *The Complete Pyramids*. Thames and Hudson, April 28th, 2008. p.20
32 Mark Lehner. *The Complete Pyramids*. Thames and Hudson, April 28th, 2008. p.24
33 Mark Lehner. *The Complete Pyramids*. Thames and Hudson, April 28th, 2008. p.35
34 E. A. Wallis Budge. *An Egyptian Hieroglyphic Dictionary*, Vol. 1. New York: Dover Publications, Inc. 1978. p.25.
35 Robert Schoch and Robert Aquinas McNally. *Pyramid Quest*. London: Tarcher/Penguin 2005. p.296.
36 Sir Flinders Petrie. *Wisdom of the Egyptians* (1940). 2019 digital edition p.71
37 The Classic Maya Ceremonial Bar. ANALES DEL INSTITUTO DE INVESTIGACIONES ESTÉTICAS, NÚM. 65, 1994, FLORA S. CLANCY University of New Mexico. p.12.
38 Vanpool, Todd & Royall, Travis & Vanpool, Christine. (2013). Archaeological Metrology: A Case Study from Paquime.
39 Vanpool, Todd & Royall, Travis & Vanpool, Christine. (2013). Archaeological Metrology: A Case Study from Paquime.
40 Morley and Renfrew 2010: O'Neill 1968. Vanpool, Todd & Royall, Travis & Vanpool, Christine. (2013). Archaeological Metrology: A Case Study from Paquime.
41 *The Archaeology of Albert Porter Pueblo (Site 5MT123): Excavations at a Great House Community Center in Southwestern Colorado*. Edited by Susan C. Ryan. 2015, p.84.
42 Hannah, Robert & Magli, Giulio. (2009). The role of the sun in the pantheon's design and meaning. Numen. 58. 10.1163/156852711X577050. https://arxiv.org/vc/arxiv/papers/0910/0910.0128v1.pdf.
43 Hannah, Robert & Magli, Giulio. (2009). The role of the sun in the pantheon's design and meaning. Numen. 58. 10.1163/156852711X577050.
44 Frank Waters. *Book of the Hopi*. Penguin Books (June 30, 1977) p.9-10.
45 Frank Waters. *Book of the Hopi*. Penguin Books (June 30, 1977) p.27.
46 https://soundsofstonehenge.wordpress.com/conclusions/ last retrieved 12/22/19.
47 https://soundsofstonehenge.wordpress.com/conclusions/ last retrieved 12/22/19.
48 Paul Devereux & Jon Wozencroft (2014). Stone Age Eyes and Ears: A Visual and Acoustic Pilot Study of Carn Menyn and Environs, Preseli, Wales, Time and Mind, 7:1, 47-70, DOI: 10.1080/1751696X.2013.860278.
49 PNAS first published February 11, 2019 https://doi.org/10.1073/pnas.1813268116.
50 https://www.sciencemag.org/news/2019/02/stonehenge-other-ancient-rock-structures-may-trace-their-origins-monuments. By Michael Price. Feb. 11, 2019, 3:00 PM.
51 Bruce Bradley and Dennis Stanford. The North Atlantic ice-edge corridor: a possible Paleolithic route to the New World. *World Archaeology* Vol. 36(4): 459 – 478 Debates in World

Archaeology # 2004 Taylor & Francis Ltd ISSN 0043-8243 print/1470-1375 online DOI: 10.1080/0043824042000303656.
52. Flinders Petrie. *Measures and Weights*. London: Methuen & Company Ltd. 1934. p. 1.
53. *The Archeology of Measurement: Comprehending Heaven, Earth and Time in Ancient Societies*. Cambridge University Press 2010. Edited by Iain Morley and Colin Renfrew. p.1.
54. Flinders Petrie. *The Journal of the Anthropological Institute of Great Britain and Ireland*, Volume VIII. London 1879.
55. C. Staniland Wake. *The Origin and Significance of the Great Pyramid*. London: Reeves and Turner, 1882.
56. Linda Schele and David Freidel. *A Forest of Kings*. Morrow (1990) p.415-416.
57. The Classic Maya Ceremonial Bar, ANALES DEL INSTITUTO DE INVESTIGACIONES ESTÉTICAS, NÚM. 65, 1994, FLORA S. CLANCY University of New Mexico.
58. Hattie Greene. *The Unwritten Literature of the Hopi*. Lockett Release Date: May 24, 2005 [EBook #15888] Online Project Gutenberg. Ch. IX HOPI MYTHS AND TRADITIONS AND SOME CEREMONIES BASED UPON THEM. Original Publication: University of Arizona TUCSON, ARIZONA: University of Arizona Bulletin. SOCIAL SCIENCE BULLETIN No. 2 Vol. IV, No. 4 May 15, 1933.
59. Leo Damrosch. *Eternity's Sunrise: The Imaginative World of William*. Yale University Press; First Edition 2015. p.194-195.
60. Kevin Duffy. *Who Were the Celts?* Barnes and Noble, Inc. 1999. P.148.
61. Manly P. Hall, *The Secret Teachings of All Ages*, Golden Anniversary Ed. Los Angeles: Philosophical Research Society, Inc., 1977. p.23.
62. Flinders Petrie. *Measures and Weights*. London: Methuen & Company Ltd. 1934. p.1.
63. Flinders Petrie. *The Journal of the Anthropological Institute of Great Britain and Ireland*, Volume VIII. London 1879.
64. Flinders Petrie. *The Journal of the Anthropological Institute of Great Britain and Ireland*, Volume VIII. London 1879:107.
65. Radiocarbon dates and Bayesian modeling support maritime diffusion model for megaliths in Europe. PNAS first published February 11, 2019. https://doi.org/10.1073/pnas.1813268116.
66. https://www.sciencemag.org/news/2019/02/stonehenge-other-ancient-rock-structures-may-trace-their-origins-monuments. By Michael Price Feb. 11, 2019 , 3:00 PM.
67. https://www.washingtonpost.com/national/health-science/radical-theory-of-first-americans-places-stone-age-europeans-in-delmarva-20000-years-ago/2012/02/28/gIQA4mriiR_story.html?utm_term=.773d632cfdcc.
68. Bruce Bradley and Dennis Stanford. The North Atlantic ice-edge corridor: a possible Paleolithic route to the New World. *World Archaeology* Vol. 36(4): 459 – 478 Debates in World Archaeology # 2004 Taylor & Francis Ltd ISSN 0043-8243 print/1470-1375 online DOI: 10.1080/0043824042000303656.
69. Posted by Current World Archaeology, March 6, 2007; https://www.world-archaeology.com/world/africa/botswana/ritual-organised-activity-identified-as-worlds-oldest/.
70. *Jung, Collected Works* vol. 8 (1960). "The Significance of Constitution and Heredity in Psychology" (1929), 229–230 (p.112).
71. *The Archeology of Measurement: Comprehending Heaven, Earth and Time in Ancient Societies*. Cambridge University Press. 2010. Edited by Iain Morley and Colin Renfrew. p.1.
72. *Plato: Complete Works* Hardcover – May 1, 1997 by Plato (Author), John M. Cooper Editor), D. S. Hutchinson (Editor) Timaeus p.1230.
73. *Plato: Complete Works* Hardcover – May 1, 1997 by Plato (Author), John M. Cooper (Editor), D. S. Hutchinson (Editor) Timaeus p.1230.
74. *Plato: Complete Works* Hardcover – May 1, 1997 by Plato (Author), John M. Cooper (Editor), D. S. Hutchinson (Editor) Critias p.1306.
75. Author: Dr. Tony Phillips | Production editor: Dr. Tony Phillips | Credit: Science@NASA.
76. https://obamawhitehouse.archives.gov/blog/2016/10/12/preparing-nation-space-weather-new-executive-order.

77 https://www.sciencemag.org/news/2018/11/massive-crater-under-greenland-s-ice-points-climate-altering-impact-time-humans By Paul Voosen Nov. 14, 2018.
78 Moore, C.R., Brooks, M.J., Goodyear, A.C. et al. Sediment Cores from White Pond, South Carolina, contain a Platinum Anomaly, Pyrogenic Carbon Peak, and Coprophilous Spore Decline at 12.8 ka. Sci Rep 9, 15121 (2019). https://doi.org/10.1038/s41598-019-51552-8 https://www.nature.com/articles/s41598-019-51552-8.
79 Pino, M., Abarzúa, A.M., Astorga, G. et al. Sedimentary record from Patagonia, southern Chile supports cosmic-impact triggering of biomass burning, climate change, and megafaunal extinctions at 12.8 ka. Sci Rep 9, 4413 (2019). https://doi.org/10.1038/s41598-018-38089-.
80 Pino, M., Abarzúa, A.M., Astorga, G. et al. Sedimentary record from Patagonia, southern Chile supports cosmic-impact triggering of biomass burning, climate change, and megafaunal extinctions at 12.8 ka. Sci Rep 9, 4413 (2019). https://doi.org/10.1038/s41598-018-38089-.
81 Hey, J. (2005). On the number of New World founders: A population genetic portrait of the peopling of the Americas. PLoS Biol 3(6): e193.
82 https://www.livescience.com/289-north-america-settled-70-people-study-concludes.html.
83 Li, H., Durbin, R. Inference of human population history from individual whole-genome sequences. Nature 475, 493–496 (2011). https://doi.org/10.1038/nature10231.
84 Pino, M., Abarzúa, A.M., Astorga, G. et al. Sedimentary record from Patagonia, southern Chile supports cosmic-impact triggering of biomass burning, climate change, and megafaunal extinctions at 12.8 ka. Sci Rep 9, 4413 (2019). https://doi.org/10.1038/s41598-018-38089-.
85 *Life After People*. History Channel documentary 2008. Director and writer: David De Vries.
86 Sweatman, M. B. and Tsikritsis, D. (2017). Decoding Göbekli Tepe with archaeoastronomy: What does the fox say? *Mediterranean Archaeology and Archaeometry* 17(1), 233-250.
87 Sweatman, M. B. and Tsikritsis, D. (2017). Decoding Göbekli Tepe with archaeoastronomy: What does the fox say? *Mediterranean Archaeology and Archaeometry* 17(1), 233-250.
88 Richard J. Reidy. *Eternal Egypt: Ancient Rituals for the Modern World*. Publisher: iUniverse (January 20, 2010) p.123.
89 Seyfzadeh, M. & Schoch, R. (2019). World's First Known Written Word at Göbekli Tepe on T-Shaped Pillar 18 Means God. Archaeological Discovery, 7, 31-53. p.35-36. https://doi.org/10.4236/ad.2019.72003.
90 Manly P. Hall. *The Secret Teachings of All Ages*. Golden Anniversary Ed. Los Angeles: Philosophical Research Society, Inc., 1977. p.23.
91 https://earthsky.org/favorite-star-patterns/the-northern-cross-backbone-of-the-milky-way.
92 Petrie. *The Journal of the Anthropological Institute of Great Britain and Ireland*, Volume VIII. London 1879: 107.
93 Thomas Keating. *The Human Condition*. Paulist Press, New Jersey. 1999.
94 George Santayana. *The Life of Reason: Reason in Common Sense*. Scribner's 1905: 284.
95 https://translate.google.com/translate?hl=en&sl=es&u=http://www.mexiconewsnetwork.com/es/arte-cultura/ichkabal-sitio-arqueologico-quintana-roo/&prev=search.
96 https://translate.google.com/translate?hl=en&sl=es&u=http://www.inah.gob.mx/zonas/106-zona-arqueologica-ichkabal&prev=search.
97 Vincent H. Malmström, *Cycles of the Sun, Mysteries of the Moon: The Calendar in Mesoamerican Civilization*. University of Texas Press. 1st edition (October 1, 1996) Chapter 3; Strange Attraction: The Mystery of Magnetism.
98 *Carl Jung, Letters*. Volume 2. Princeton University Press, 1973. p.33. https://libquotes.com/carl-jung/quote/lbi7w9m.
99 A publication of the German research project Mohenjo-Daro, Department for history of Architecture and Architectural preservation, RWTH Aachen, IsMEO, Rome in cooperation with the archeological survey of Pakistan. 1984